Essence of
VEGETABLE SCIENCE

The Authors

Dr. Manish Kumar Singh is an Assistant Professor in the Department of Vegetable Science, College of Horticulture, Banda University of Agriculture and Technology, Banda, Uttar Pradesh, India and author or co-author of more than 40 peer-reviewed journal articles, 1 Book, 8 book chapters, and several conference papers. Recently, Dr. Singh is conferred the Young Scientist Award by the BRIATS, Allahabad, Best Poster presentation award by Uttar Pradesh Council of Agricultural Research, Lucknow, Best Article award by Agriculture and food e-news *etc.* He is also handling 1 project as PI and 4 project as Co PI funded by different agencies. Dr. Singh acquired B.Sc. (Ag.) M. Sc. (Ag.) in Horticulture and Ph. D. Horticulture from Banaras Hindu University, Varanasi, India with specialization in Vegetable Science. During the Ph D. research he has been awarded as UGC Scholarship as well as qualified ASRB-NET

Dr. Amit Kumar Singh is an Assistant Professor in the Department of Post-Harvest Technology, College of Horticulture, Banda University of Agriculture and Technology, Banda, Uttar Pradesh, India and author or co-author of more than 25 peer-reviewed journal articles, 7 Manuals, 3 Books, 11 book chapters, and several conference papers. Recently, Dr. Singh is conferred the Research Excellence Award-2022 for outstanding research work by SCNR, Kolkata, the Young Scientist Award- 2018 by the BRIATS, Allahabad, the Young Associated Scientist Award- 2017 by the Agricultural Technology Development Society, Ghaziabad, Fellow International College of Nutrition-2016 *etc.* He has been an active member of the organizing committee of several national seminars, webinars and brain storming sessions *etc.* Dr. Singh acquired B.Sc. (Agriculture) degree from Post Graduate College, Ghazipur, India. He received the M. Sc. (Horticulture) and Ph. D. (Horticulture) degrees from Bidhan Chandra Krishi Viswavidyalaya, Mohanpur, Nadia, India with specialization in Post-Harvest Technology

Dr. Sudhir Kumar Mishra is an Assistant Professor in the Department of Horticulture, Faculty of Agriculture, National Post Graduate College, Barahalganj, Uttar Pradesh, India and author or co-author of more than 30 peer-reviewed journal articles, 1 Book, and several conference papers. Recently, Dr. Mishra is conferred the Young Extension Worker Award by the Bihar Veterinary College, Patna, Best Poster presentation award, Best Paper presentation Award *etc.* Dr. Mishra acquired B.Sc. (Hort.) M. Sc. (Hort.) in Horticulture and perusing Ph. D. Horticulture from Gorakhpur University, Gorakhpur. During the Ph. D. research he has been qualified ASRB-NET .

Essence of VEGETABLE SCIENCE

Dr. Manish Kumar Singh

Assistant Professor, Department of Vegetable Science
College of Horticulture,
Banda University of Agriculture and Technology, Banda, Uttar Pradesh

Dr. Amit Kumar Singh

Assistant Professor, Department of Post-Harvest Technology
College of Horticulture,
Banda University of Agriculture and Technology, Banda, Uttar Pradesh

Dr. Sudhir Kumar Mishra

Assistant Professor, Department of Horticulture, Faculty of Agriculture,
National Post Graduate College, Barahalganj, Uttar Pradesh

2024

Astral Cracker
A Division of
Astral International Pvt. Ltd.
New Delhi – 110 002

ISBN: 978-93-5461-975-5

Published by : **Astral Cracker**
A Division of
Astral International Pvt. Ltd.
– ISO 9001:2015 Certified Company –
4736/23, Ansari Road, Darya Ganj,
New Delhi-110 002
Ph. 011-43549197, 23278134
E-mail: info@astralint.com
Website: www.astralint.com

Printed at : **Replika Press Pvt. Ltd.**

Vice Chancellor Office

Banda University of Agriculture & Technology, Banda-210001, U.P. Tele Fax No:- 05192-232305

Ph-05192-232305; website: buat.edu.in; *email-vc.buat@gmail.com*, *vcbuat.bn-up@gov.in*

Ref.:- VC/2023/134 **Date: 14/06/2023**

Foreword

Horticulture sector become one of the major drivers of growth in the agriculture sector and it provides employment opportunities across primary, secondary, and tertiary sectors, particularly vegetable crops are relatively resilient to changes in weather conditions. At present, the total horticultural production has overtaken as compared to the total food grains in the country. The percentage share of horticulture crops in value of total agriculture output is around thirty percent. It is a matter of pride that India is the second-largest producer of vegetables in the world. According to FAO (2021), India is the largest producer of ginger and okra among vegetables and ranks second in the production of Potatoes, Onions, Cauliflowers, Brinjals, Cabbages, etc. and the major destinations for Indian fresh vegetables are Bangladesh, UAE, Nepal, Netherland, Malaysia, Sri Lanka, the UK, Oman, and Qatar. In view of the nutritional and commercial importance of such vegetable crops, various effects on research and development of crop technologies with respect to crop improvement, crop husbandry, crop production, post-harvest handling, and value addition at Central Institutes, Regional Laboratories, and State level institutes have been made and highlighted towards the growers.

It is so gratifying to note that the contributors of the book have come forward to bring a comprehensive review and research note on these commercially important vegetable crops. The book will be very useful to the faculties, scientists, researchers, students, and industry persons and also a reference for the education personnel.

I compliment the authors, Dr. Manish Kumar Singh, Dr. Amit Kumar Singh, and Dr. Sudhir Mishra for making a sincere effort for bringing out this useful publication "Essence of Vegetable Science".

(Narendra Pratap Singh)
Vice Chancellor

भाकृअनुप
ICAR

डॉ. ऊधम सिंह गौतम
उप महानिदेशक (कृषि विस्तार)
Dr. U.S. Gautam
Deputy Director General (Agril. Extn.)

भारतीय कृषि अनुसंधान परिषद
कृषि अनुसंधान भवन–1, पूसा, नई दिल्ली 110 012
INDIAN COUNCIL OF AGRICULTURAL RESEARCH
Krishi Anusandhan Bhawan-I, Pusa, New Delhi – 110012
Phone: 91-11-25843277 (O)
E-mail: ddg-extn.icar@gov.in ; us.gautam@icar.gov.in

FOREWORD

Agricultural education is passing through a metamorphic stage of development, which hinges upon teaching, research, and extension interface. Having achieved the tremendous progress in research and technology transfer, the focus of planning is now on Human Resource Development and Human Resource Management in agriculture. Future generation of learners in agriculture will have to compete in a totally new environment having the characteristics of globalization, internationalization, innovation, and skills development. In these new paradigms, this book has been designed to facilitate scientists, academicians, extension workers as well as the stakeholders to upgrade their knowledge in the field of vegetable sciences. This book has focused on the major vegetables (Potato, Tomato, Chilli, Brinjal, Cauliflower, Cabbage, Cucumber, Musk melon, Watermelon, Pumpkin, Bottle gourd, Bitter gourd, Sponge gourd, Okra, Pea, Cowpea, Spinach, Amaranthus, Beetroot, Carrot, Radish, Onion, Garlic, Sweet potato, Tapioca, Elephant-foot-yam), minor vegetables (Knol-khol, Broccoli, Brussels sprout, Lettuce, Celery, Ridge gourd, Moringa, Curry leaf) and underutilized vegetable crops (Horse radish, Spine gourd, Chenopodium, Chinese potato, Globe artichoke, Colocassia, Tannia, Yam, Yam bean, Winged bean, Lima bean, Ash gourd, Ivy gourd, Chow-chow, Snake gourd, Pointed gourd) for their production, improvement, protection, and processing technology, to those who are preparing for ICAR-AIEEA, JRF, SRF, NET, ARS, CUET, UPCATET examinations.

The book entitled "Essence of Vegetable Science" is a valuable addition to the upcoming collection for knowledge upgradation of under-graduate and post-graduate students. I appreciate the commendable efforts made by Dr. Manish Kumar Singh, Dr. Amit Kumar Singh, and Mr. Sudhir Mishra in editing this publication. The academic contributions made by different faculties form the strong backbone of the publication. I am sure that this publication will be an important source-book for the stakeholders in vegetable science.

(Udham Singh Gautam)
DDG Extension (ICAR)
& Former VC - BUAT, Banda

Preface

India enjoys a prominent position on the horticultural map of the world. The varying weather conditions of this country provide a suitable environment for growing a variety of vegetables. Globally, India is a leading vegetable-producing country with a share of 15 per cent, however sufficient opportunity exists to improve its productivity. Due to a wide range of their productivity vegetable growers earn more profit than growing other commodities. Presently horticultural production suppressed food grain production with a record vegetable production of 204 million tonnes. Vegetable production is considered an important component of crop diversification and plays a crucial role in food, nutritional, economic, and livelihood security. It constitutes the most important part of food and provides critical nutrients for a balanced diet.

On this deliberation, the present book has been attempted which has pursued one vital part in the area of production, improvement, and protection as well as their post-harvest management of different vegetables (Solanaceous crops, Cole crops, Cucurbitaceous crops, Leguminous crops, Root and Tuber crops, Leafy vegetables, Okra, Drumstick, Underutilized vegetables *etc.*). The book thus; seeks to meet the requirements of the undergraduate and post-graduate students of vegetable science, horticulture, post-harvest technology and related areas on the one hand and those who desire to take up vegetable production and their post-harvest management in the other hand. We are sure that this book will be very useful to the students who are preparing for JRF, SRF, ARS, NET, *etc.* in vegetable science and post-harvest management.

In this book, published information has been freely used. Some information has been collected from electronic media/websites, the sources of which have duly been acknowledged. We would like to record our special thanks to the Chancellor and Vice-Chancellor of the Banda University of Agriculture and Technology, Banda to encourage and facilitate for the making of this valuable book. Lastly, the author's looks forward to receiving valued opinions and suggestions from the readers so as to bring out the book in a better form in the future.

Manish Kumar Singh
Amit Kumar Singh
Sudhir Kumar Mishra

Contents

Chapter 1

Solanaceous Vegetables

Tomato

- *Solanum lycopersicum* L.; Solanaceae, 2n=2x=24,Origin: -South America (Mexico)
- Common name- Love of apple (England), Poor man's orange in (India), Wolf apple, Vilayati Baigan.
- Fruit type is berry and flower type is hermaphrodite (bisexual)
- Father of tomato breeding is Dr. C.M Rick
- Father of tomato in India is Dr. G.Kallo
- Tomato genome size= 950mbp
- Tomato genome was fully sequenced in 2012
- Tomato seeds contain oil - 24 per cent
- Tomato pigments-Red colour-lycopene (at 20-25 ^{0}C)
- Tomato introduced in India in 19th century
- Test weight of tomato seed- (2.5-3.3 gm)
- No of stamen-(6)
- Isolation distance (50 meter for foundation seed and 25 meter for certified seeds)
- Yellow carotenoids pigment is due to β-carotene
- Tangerine colour is due to pro-lycopene pigment
- Citric acid contains is 0.3 per cent

- ☆ Toxic element is tomatine and saponine
- ☆ Lycopene contains in tomato fruits-25-50mg/100gm.
- ☆ Type of Inflorescence – Racemose (Cyme)
- ☆ Type of fruit-Multi seeded berry
- ☆ Highly self-polineted crops due to chasmogamy and homomorphism
- ☆ Optimum temperature for pollination- 21 ^{0}C.
- ☆ Tomato colour determined by colorimeter
- ☆ Fruit type (berry), cluster of flowers (Truss)
- ☆ It's a self-pollinated $C_{3,}$ and Day neutral plant
- ☆ Flowers are perfect and hypogynous
- ☆ Cherry tomato mainly used in Europe country
- ☆ Tomato seed contain 24 per cent oil used in the manufacture of margarine.
- ☆ Vit. A present in tomato 320mg.
- ☆ Tomatine contains small and hard fruit -300mg and large green tomato -150mg.
- ☆ Tomatine is found more in amount in leaves/flowers
- ☆ Tomatine is aalkaloid, and found in tomato leaves, which is non-toxic in nature.
- ☆ Mullar and luckwill divided the genous (*Lycopersicon*) into two sub - genera
- ☆ Eulycopersicon- Red fruited tomatoes (carotenoied pigment)
- ☆ Eriolycopersicon -Green fruited tomatoes (anthocyanin pigment)

Plants are two type –

- ☆ Determinate type –Inflorescence occurs at every internode, self topping in nature.
- ☆ Indeterminate type- Inflorescence occurs at every 3rdinternode, required stacking.
- ☆ Training and pruning are followed in indeterminate type tomato
- ☆ Staking is followed in indeterminate type tomato
- ☆ Total chromosome pair- 12
- ☆ Chromosome-1 is largest
- ☆ Chromosome-2 is (satelite chromosome)
- ☆ Chromosome-12 is (shortest chromosome)
- ☆ The genous (*lycopersicon*) includes (12 species).

- ✰ Tomato has divided into (5 forms): Cherry tomato (*L. cerasiforme*); Pear tomato (*L. pyriformae*); Common tomato (*L. commune*); Potato leaf tomato (*L. grandifolium*);Upright leaf tomato (*L.validium*)
- ✰ BIS grade-(super A, super, fancy, commercial)
- ✰ Genetic modified tomato – Resistant to alzheimers disease.
- ✰ Parthenocarpic fruit is controlled by gene (*pat -2*)
- ✰ Small fruit genotype of tomato is rich in ascorbic acid.
- ✰ Optimum temperature for pollination is 21 ^{0}C
- ✰ Day temperature of (26.5 ^{0}C) and night temperature between (16-22 ^{0}C) resulted heaviest fruiting.
- ✰ Citric and malic acids are the main acid found in tomato,but citric acid is the major constituent of total acid.
- ✰ β –carotene –(increase mature green to ripening stage).
- ✰ Red –lycopene pigment, yellow – carotenoid pigment.
- ✰ Tangerine –poly-lycopene pigment
- ✰ Temperature –effect on fruit colour development
- ✰ Bellow 10 ^{0}C red or yellow colour development stop.
- ✰ Above 30 ^{0}C red colour suppressed.
- ✰ Above 40 ^{0}C lycopene colour destroyed and no red colour developed.
- ✰ Best temperature for lycopene development (21-24 ^{0}C)
- ✰ Red light enhance (chlorophyll degradation).
- ✰ Wild tomato contained larger amount of tomatine.
- ✰ Tomato fruits are failed to set seed at (13 ^{0}C) or bellow
- ✰ Optimum temperature ranges for fruit set (15-20 ^{0}C)
- ✰ Hardening of transplanting seedling can't be done by using (0.2 per cent CCC).
- ✰ TSS, sugar, protein, ascorbic acid, increase by spray kinetin @100ppm
- ✰ Higher price harvesting of tomato fruit – Pink stage.
- ✰ Tomato are harvested for markets stage– Pink stage
- ✰ Poleries harvester used for tomato harvester developed by –G.D.GHANNA
- ✰ Self-life of tomatoes fruit is maximum at 21^{0}C.
- ✰ *NOR* (not over ripening) type fruit had the best storage for self-life
- ✰ Early blight of tomato was 1st recorded (1882 in new jersy)
- ✰ Root stock –KNVS (*L. esculentum X L. hirsutum*)

- ☆ Normal red fruit ratio of lycopene: caroteneis 12-15: 1.
- ☆ Fruit colour is controlled by 3 genes independent (B,R,T)
- ☆ Wild tomato –*L. peruvianum* (warm season crops)
- ☆ Flavour saver – It is a transgenic tomato 1st develop in 1994
- ☆ Endless summer-It is a transgenic var. using (*ACC* synthesis activity)
- ☆ Male sterility in tomato was 1st reported by Rick (1944)
- ☆ Larson and pour (1948) functional male sterility mutant tomato: john baervar.
- ☆ Soil acidity moderately tolerant at 5.5-6.8 pH
- ☆ Soil salinity moderate and water requirement moderate.
- ☆ Root depth- very deep
- ☆ Tomato unripe storage 4-5 weeks at temperature 8.9-10 ^{0}C and 85-90 per cent RH
- ☆ Ripe tomato storage- one week at temp 7.2 ^{0}C and 90 per cent RH
- ☆ Phosphorus improved – Root development
- ☆ Ancestor of tomato-(*Lycopersicum esculentum* var. *ceresiform*)
- ☆ Toxic substance- Saponine.
- ☆ Sex form –chasmogamy.
- ☆ Climacteric vegetable crops.
- ☆ Golden tomato developed by AVRDC, Taiwan
- ☆ 3-6 times β-carotene higher in golden tomato
- ☆ Ca deficiency (blossom end rot- BER)
- ☆ Tomato genetic resource centre (TGRC) Devis (USA)
- ☆ International solanaceous genome centre project –Cornel University (USA)
- ☆ Effect of ethylene increase (lycopene and carotene)
- ☆ Vitamin-C- (31mg/gm)
- ☆ 33 per cent F_1 tomato seeds used in cultivation
- ☆ Tomato fruit aroma (sulfonium)
- ☆ Tomato seed germination inhibited compound is caffic and ferulic acid.
- ☆ Best method of seed extraction is alkali treatment method uses 10 per cent washing soda.
- ☆ Acid treatment method uses HCL @ 250ml/10 litre for 10 minutes

- ☆ Tomato juice contain 5 per cent TSS, Tomato sauce-30 per cent TSS and 1.2 per cent acidity, chutney (50 per cent TSS)
- ☆ Tomato paste-(25 per cent TSS),Ketchup-(28 per cent TSS), Tomato puree(9 per cent TSS)
- ☆ Tomato soup-(7 per cent TSS)
- ☆ Pickle and Chateny 100 PPM- SO_2 and 250 PPM Benzoic Acid/Sodium Benzoate
- ☆ Dehydrated vegetable –(2000 PPM SO_2)
- ☆ Avg. Seed yields (100-120 kg/ha).
- ☆ Tomato yield - 250-400q/ha.
- ☆ Seed rate (125-175gm/ha.)
- ☆ High β-carotene *var.* Caro rich and caro red).

Varieties

- ☆ Parthenocarpy (*severnine*).
- ☆ Var. Pusa Sadabahar is best suitable for cultivation in high and low temperature
- ☆ Processing var. Roma, S-152, Pusa Sheetal, Pant Bahar, Pusa Uphar
- ☆ Hilly var. Sioux, Best of all (IARI) Marglobe.
- ☆ Plain area- Pusa Ruby, Pusa Early Dwarf, Pusa-120
- ☆ Indeterminate var. Pusa Ruby, Arka Saurabh, Arka Vikash, Best of all, Pusa Divya, Arka Abhijit.
- ☆ Juicy quality-NDT 120, ketchup - Pant Bahar (fuserium wilt resist. Var.)
- ☆ Pusa Gaurav-(Glamour X Watch)- Suitable for canning
- ☆ Pusa Sheetal-fruit set at low temp. (8 ^{0}C), chutney making var.-NDT-101
- ☆ High density planting var.-(processor 278), extreme early var. Hissar Arun
- ☆ Determinate type var.- Pusa Sheetal, Panjab Chuhara, Roma
- ☆ Highest acidity hybrid var.-(KT-4), Nematode resist. Var.-(Arka Vardan).
- ☆ Open pollinated var.-(Indam-13, Indam-15)
- ☆ Fusarium wilt resistant var.-Marglobe, Pritchard
- ☆ Root knot nematode resistant var.-Nematox, Pusa-120, Hissar Lallit.
- ☆ Bacterial resistant var.-Arka Abhijit, Arka Alok, Arka Shrestha
- ☆ Pusa Ruby -(Sioux X Improved Meeruti).
- ☆ Pusa-early dwarf (Improved Meeruti X Red Cloud) best combiner variety.
- ☆ Panjab Chhuhara-(EC-55005 X Panjab Tropic)

- Pusa Red Plum-(*L. esculentum X L. pimpinelifolium*)
- 1st private sector commercial var.-(karnatka-1) in 1973, (Pusa Divya- F1 hybrid var.)
- Pusa Divya –(Long style Psiang X Roma)
- Sweet-72- (Pusa Red Plum X Sioux)
- S-152, Pusa Gaurav (all-purpose var.)
- Inter-spacific var. (Hisar Anmol, Pusa Red Plum)
- Hisar Anmol-(Hisar Arun X *L. hirsutam*)
- Leaf curl resistant var.- (Hisar Gaurav and Hissar Anmol)
- Long transport var.-Pusa Rohini, Punjab Chhuhara
- Arka Meghali- (rainfed condition), Arka Vikash –(drought condition)
- Triple resistant var. Arka Rakshat, mutant var.(S-12, PKM-1, Maruthan, Pusa Lal Meeruti)
- RIN type suitable for- (cold storage)
- 90 per cent yield loss to TLCV
- Soil: Well drained loamy soil rich in organic matter with a pH range of 6.5 - 7.5.
- Season of sowing is May - June and November - December
- Recommended dose of fertilizers is FYM 10 kg, Neem cake 1 kg, VAM 50 g, enriched Superphosphate 100 g and Furadon 10 g per square metre before sowing.
- Seed rate Varieties: 300-350 g/ha Hybrids: 100-150 g/ha
- Seed treatment: Treat the seeds with *Trichoderma viride* 4 g or *Pseudomonas fluorescens* 10 g or Carbendazim 2 g per kg of seeds 24 hours before sowing.
- Just before sowing, treat the seeds with *Azospirillum* @ 40 g/400 g of seeds. Sow in lines at 10 cm apart in raised nursery beds and cover with sand.
- Protected nursery: Prepare the nursery area of 3 percents with slanting slope of 2 per cent for the seedling production to cover 1 ha.
- Cover the nursery area with 50 per cent shade net and cover the sides using 40/50 mesh insect proof nylon net.
- Form raised beds of 1 m width and convenient length and place HDPV pipes at 2 m interval for further protection with polythene sheets during rainy season.

- Mix sterilized coco-peat @ 300 kg with 5 kg neem cake along with *Azospirillum* and Phospho-bacteria each @ 1 kg; approximately 1.2 kg of coco-peat is required for filling one pro-tray.
- 238 pro-trays (98 cells) are required for the production of 23,334 seedlings, which are required for one hectare adopting a spacing of 90 x 60 x 60 cm in paired row system.
- Sown the treated seed in pro-trays @ one seed per cell.
- Cover the seed with coco-peat and keep the trays one above the other and cover with a polythene sheet till germination starts
- After six days, place the pro-trays with germinated seeds individually on the raised beds inside the shade net
- Water with rosecanevery day and drench with NPK 19: 19: 19 @ 0.5 per cent (5g/l) at 18 days after sowing
- Field preparation: Plough the land to fine tilth. Thoroughly prepare the field with the addition of FYM @ 25 t/ha and form ridges and furrows at a spacing of 60 cm. Apply 2 kg/ha of *Azospirillum* and 2 kg/ha of Phosphobacteria by mixing with 50 kg of FYM.
- Irrigate the furrows and transplant 25 days old seedlings on the sides of ridges.
- Life irrigation to be given on 3rd day of planting.
- Mulching - Mulch with black LDPE sheets of 25-micron thickness and bury both the ends into the soil to a depth of 10 cm
- Weed control – Apply Pendimethalin 1.0 kg a.i./ha or Fluchloralin 1.0 kg a.i./ha as pre-emergence herbicide, followed by hand weeding once at 30 days after planting.
- Irrigation- After establishment of seedlings, irrigate at weekly intervals.
- Apply FYM @ 25 t/ha as basal before last ploughing.
- Apply 2 kg/ha of *Azospirillum* and 2 kg/ha Phospho-bacteria by mixing with 50 kg of FYM.
- Apply 75 per cent total recommended dose of superphosphate i.e 1172 kg/ha as basal.
- Install the drip irrigation with main and sub main pipes and place lateral tubes at an interval of 1.5 m. Place the drippers in lateral tubes at an interval of 60 cm and 50 cm spacing with 4 LPH and 3.5 LPH capacities respectively. Form raised beds of 120 cm width at an interval of 30 cm and place the laterals at the centre of each bed.
- Before planting, wet the beds using drip system for 8-12 hrs.

- ✫ Spray Pendimethalin 1.0 kg a.i./ha or Fluchloralin 1.0 kg a.i/ha as pre-emergence herbicide at 3rd day after planting.
- ✫ Gap filling has to be done at 7th day after transplanting.
- ✫ Basal dose: FYM 25 t/ha, NPK 75: 100: 50 kg/ha
- ✫ Borax 10 kg and Zinc sulphate 50 kg/ha
- ✫ Top dressing: 75 kg N/ha on 30th day of planting or during earthing up.
- ✫ Hybrids Basal dose - FYM 25 t/ha, NPK 50: 250: 100 kg/ha
- ✫ Borax 10 kg and Zinc sulphate 50 kg/ha
- ✫ Top dressing: N and K each 150 kg/ha in 3 equal splits at 30, 45 and 60 days after planting.
- ✫ Growth regulators: Spray 1.25 ppm (625 ml in 500 litres of water) Triacontanol at 15 days after transplanting and at full bloom stage to increase the yield.
- ✫ Training of hybrids: Stake the plants 30 days after planting with 1 - 1.5 m tall stakes.
- ✫ Remove the side branches up to 20 cm from ground level.
- ✫ Micronutrient spray: Foliar spray of ZnSO4 @ 0.5 per cent thrice at 10 days interval from 40 days after planting.
- ✫ Spray 19: 19: 19 + Mn @ 1 per cent at 60 days after planting.
- ✫ Protected cultivation of tomato in polyhouses: During rainy season, the indeterminate tomato hybrid has to be grown in the medium consisting of FYM: composted coir pith: sand (2: 1: 1) with irrigation regime of 20k Pa and basal application of 50kg each of NPK/ha as straight fertilizers and 250 kg each of NPK as water soluble and straight fertilizers through fertigation along with black polyethylene mulch (50 microns)
- ✫ **Fruit borer:** *Helicoverpa armigera* and *Spodoptera litura* (common for both)
- ✫ Grow simultaneously 40 days old American tall marigold and 25 days old tomato seedlings @ 1: 16 rows.
- ✫ Set up pheromone traps @ 12/ha.
- ✫ Collect and destroy damaged fruits and grown-up caterpillars.
- ✫ Release *Trichogramma pretiosum* @ 1 lakh/ha/release at an interval of 7 days starting from flower initiation stage.
- ✫ Release *Trichogramma chilonis* @ 50,000/ha/release, coinciding with flowering time and based on ETL of 10 per cent damage.

- For *Helicoverpa armigera*: H.a. NPV 1.5 x 1012 POBs/ha i.e. ***NPV*** of *H.armigera* 0.43 per cent AS @ 3.0 ml/l or 2 per cent AS @ 1.0 ml per l For *Spodoptera litura*: S.l. ***NPV*** 1.5 x 1012 POBs/ha.
- **Serpentine leaf miner**: Spray Neem Seed Kernel Extract 5 per cent.
- **Whitefly:** Install yellow sticky traps @ 12/ha to attract the adult. Remove alternate weed host Abutilon indicum. Apply carbofuran 3G @ 40 kg/ha or spray any one of the following insecticides
- **Nematode:** Soil application of *Bacillus subtilis* (BbV 57) or *Pseudomonas fluorescens* as seed treatment @ 10 g/kg of seeds and soil application (SA) @ 2.5 kg/ha for the management root knot and reniform nematode infestation in soil and root.
- Application of liquid formulation of *Bacillus subtilis* (BbV 57) or *Pseudomonas fluorescens* @ 1000 ml/ha through drip irrigation for the management of root knot nematode in tomato.
- Damping off (nursery): Treat the seeds with Trichoderma viride @ 4 g/kg or *Pseudomonas fluorescens* @ 10 g/kg of seed 24 hours before sowing.
- **Leaf spot:** Spray Zineb or Mancozeb @ 2 g/l.
- **Fusariaum wilt and Root knot nematode**: Soil solarization before preparation of nursery bed.
- **Leaf curl**: Spray systemic insecticides like Methyl dimeton or Dimethoate at 2 ml/l to kill the insect vector, whitefly.
- **Tomato spotted wilt virus**: Carbofuran 3G @ 33 kg/ha in nursery at sowing and second application @ 40 kg/ha on 10 days after transplanting in main field
- **Peanut bud necrosis virus**: Selection of healthy seedlings and rouging of PBNV infected plants up to 45 days of planting.
- Duration: 110- 115 days from transplanting (135 - 140 days from sowing)
- Rouging out of virus infected plants up to 45 days of transplanting Grow marigold as a border crop
- Set up ***Helicoverpa/Spodoptera*** pheromone traps @ 12 numbers/ ha.Release Release *Trichogrammachilonis* @ 50000/ha. Install yellow sticky traps
- Spraying Neem formulations (1 per cent) or Neem seed kernel extract (5 per cent)

Brinjal

- Discolouration of brinjal is due to polyphenol oxidase (PPO)
- Dry fruit of brinjal is due to goitrogenic principle
- Botanical name- ***Solanum melongena***
- Chromosome number - 2n=24
- Origin- India
- Common name- egg plant/aubergine/poor manscurf
- Day neutral plant
- Type of inflorescence- Cyme
- Tolerance to soil salinity- Less tolerant
- Root depth- 60-90 cm
- Climactric vegetable
- Seed rate- 200-300 gm/ha, hybrid- 150-200gm/ha.
- 1 gm seed of brinjal= 250 seeds
- Toxic substance solosodine harmfull substance
- White brinjal is good for diabetic patient
- India is 2ndlargest producer in world after china
- Bitter taste in brinjal is due to gyloco alkaloids solosodine
- De- cholestrolling action is due to the presence of poly- unsaturated fatty acid which are present in flesh and seeds and fruit in higher plant
- Highest producing in brinjal Karnataka -22.1tonnes/ha
- National average productivity – 16ton/ha
- No- inbreeding depression brinjal- Pentamorous flower
- Aroma compound- Nasunin (anthocynin)
- Most favoured temp. of brinjal production- 21-27 ^{0}C
- Optimum temperature of brinjal seed germination 20-25^{0}C with R.H 50 per cent
- Seedling are transplanted best when temp. is around 20^{0}C
- Heterostyle is common feature in multi flower except in bunchy type
- Storage temperature 10-11 ^{0}C R.H- 92 per cent
- Storage life - 3-4 weeks
- Male sterility- Genetic male sterilitsssy

- ☆ Max. fruits setting in long style flower 70-86 per cent
- ☆ Brinjal is a self-pollinated crop
- ☆ Low temp. in once more short style flower
- ☆ Low temp. leads to ovary splititing
- ☆ Brinjal cover 8-14 per cent of total area 9.0 per cent of total veg. production
- ☆ Anthocynine pigment present in fruit is reposible for purple colour of fruit skin
- ☆ Proginator of cultivated brinjal ***solanum incanum***
- ☆ African brinjal- ***S. macrocarpan***
- ☆ Immune to RKN root knot nematodes- *S. sisymbrifolium* × *S. indicum*
- ☆ Resistant to shoot and fruit borer- *S.khasianum*× *S. viarum*
- ☆ Resistant to phemopsis blight- *S. xanthocarpan*
- ☆ Little leaf disease-*S. auriculatum* × *S. gilo*
- ☆ Green brinjal self life 4 weeks
- ☆ **Nagai and kida- 1st hybrid 1926**
- ☆ 2-4D @2ppm flowering induce parthenocarpy, increase fruit set
- ☆ Major pest of brinjal- shoot and fruit borer (*lencinodes orbonalius*)
- ☆ 70 per cent yield loss due to shoot and fruit borer
- ☆ Transgenic brinjal in India develop by MAHYCO using *cry 1 AC* gene
- ☆ Serious disease in seed production- Phomopsis blight
- ☆ 1939 1st reported in little leaf disease – **Thomas/krishnawami**
- ☆ Little leaf transmitted by leaf hopper (**jassids)**
- ☆ Little leaf of brinjal causing MLO- loss in 40-80 per cent
- ☆ Resistant to fusarium wilt- florida market
- ☆ Suited for ratoon cropping- Hisar Jamuni
- ☆ Bacterial wilt resistance var.- Pusa purple/cluster/aromatic
- ☆ Resistance to phomophsis fruit rot- Pusa bhairav/Pusa Anupam
- ☆ Soft joint and easy to harvest- NDB-25
- ☆ KKM-1, white colour/suitable for diabetes patent
- ☆ PKM-1 suitable long distance transport
- ☆ Resistant to bacterial wilt- Arka Neelkanth (BWR-54)
- ☆ Pusa Purple Long- resistance to shoot and fruit borer
- ☆ Pusa Anamol- PPL× Haiderpur

- ☆ Pant rituraj- Both season var.
- ☆ Annamalai- Aphid resistant var.
- ☆ Cu content and polyphenol oxidase activity are highest in purple colour fruit and lowest in white colour
- ☆ Round type of brinjal have higher PPO
- ☆ Anthocynine pigment present in brinjal
- ☆ Flower- Hermaphrodite
- ☆ Green colour var.- Arka Shirish/Arka Kusumakar
- ☆ Extra early maturity var.- Pusa purple long
- ☆ Thornless var.- Punjab bahar, Azad Kranti (1983)
- ☆ Seedless- Arka Satish
- ☆ Amino acid content is higher in the purple fruited cultivar.
- ☆ Lowest in white colour
- ☆ Brinjal flowers are pentamorous
- ☆ Round shape/egg plant- *S. melongena var. esculentum*
- ☆ Long slender type- *S. melongena var.serpentinum*
- ☆ Dwarf brinjal- *S. melongena var.depressum*
- ☆ Chance of cross pollination is more in long style flower
- ☆ Fruit setting in long style flower varied from 70-86 per cent
- ☆ Sowing time- North India, June-july
- ☆ Seedling are ready in 4 weeks for transplanting
- ☆ Hill areas-March-April
- ☆ Spacing- 60-45, 60-60,75-60,75-75cm
- ☆ 1 per cent urea increase in fruit set and growth
- ☆ Chemical weed control- fluchloralin 1-1.5kg/ha
- ☆ Drip irrigation is most
- ☆ Seed yield 1500-2000 kg/ha
- ☆ Early crop- 20-30 tonnes/ha
- ☆ Long duration- 35-40 tonnes/ha
- ☆ F1 hybrid- 45-90tonnes/ha
- ☆ Opening of flower 6-7.30am in summer, 11.15 am in winter season
- ☆ Zn element improper of weight
- ☆ Cu, Mn increases the no. of flower fruit

- ☆ Suitable for bharta- Dudhia/Bh-2
- ☆ Pusa Anupam- Pusa Kranti × PPC- long type
- ☆ Punjab Neelam- Jamunigole × Pant Kranti
- ☆ Panjab sadabahar- good for ratoon crop
- ☆ Nematodes resistant var. black beauty/banarasjoint/vijay
- ☆ Arkashiris- hadda beetle resistant
- ☆ Private sectore- var.dan/Nisha/shiva/suphal/vaisali
- P Phomopsis blight and bacterial wilt- Pant Samrat
- ☆ Round type- Pant rituraj- t3×PPC
- ☆ Pusabarsati- PPL×H-4
- ☆ Pusabindu- SR×Rituraj
- ☆ Panjab chamkila- selection var.
- ☆ Pusa Kranti- PPL×Haiderpur×Wynard local
- ☆ Mimosa pudica increase insect activity
- ☆ Pusa Ankur- new variety
- ☆ Yield increase by Cu application in brinjal
- ☆ Fertile hybrid- *solanum incanum× S.melongena*
- ☆ Phomosphis blight resistant var.- Pusabhairav/pusa Anupam/florida market
- ☆ Bacterial resistant var. Pusa Anupam/Arka Nidhi/Arka Neelkanta/Pant Rituraj/Arka Keshav/Pusa purple cluster
- ☆ Pusa Anmol- PPL×Haiderpur
- ☆ Arka Navneet- round type- IIHR221×Supreme- high yield
- ☆ Hissajamuni- suitable for ratoon cropping- aushey×BR-34
- ☆ Anthesis time- 6-8am in August
- ☆ 9.30-11.15 am in winter
- ☆ Arka Anand- resist suitable for rabi and kharif and bacterial wilt
- ☆ Kashi Sandesh, Kashipraksh, Kashikomal, Arkanidhi, Arka Keshav, Arka Neelkant
- ☆ Purusauttam, Pusa Upkar, Pusa Aankur, Hissar Pragati
- ☆ Yello fruited spp- Saleman ferox
- ☆ 1st reported heterosis- **Kakizaki 1931**
- ☆ 1st moleculer map published by nuhome at. all- 1998
- ☆ 1st male sterility was reported by- Jasmine-1954

- ☆ 1st functional male sterility line- UGA1 MS cv. Florida high bush
- ☆ Bt. brinjal resist to shoot and fruit borer *cry-1 AC* gene
- ☆ Phomopsis blight resist governed by- **Polygene recessive**
- ☆ BR-112, PBR91-2, (202-9), (120-6)- Use for hybrid seed production
- ☆ Soil: Well-drained soil rich in organic matter with pH of 6.5-7.5.
- ☆ Nursery bed preparation Apply FYM 10kg, neem cake 1kg, VAM 50g, enriched super phosphate 100 g and furadon 10g per square metre before sowing.
- ☆ Area required for raising seedlings for planting 1.0 ha is 100 sq.m.
- ☆ Seed treatment: Treat the seeds with *Trichoderma viride* @ 4 g/kg or *Pseudomonas fluorescens* @ 10 g/kg of seed.
- ☆ In raised nursery beds, sow the seeds in lines 10 cm apart and cover them with sand.
- ☆ Transplant the seedlings 30 – 35 days after sowing at 60 cm apart in the ridges.
- ☆ Prepare the nursery area of 3 cents with a slanting slope of 2 per cent for the seedling production to cover 1 ha.
- ☆ Cover the nursery area with a 50 per cent shade net and cover the sides using a 40/50 mesh insect-proof nylon net.
- ☆ Sow the treated seeds in pro-trays @ 1 seed per cell.
- ☆ Cover the seed with coco-peat and keep the trays one above the other and cover with a polythene sheet till germination starts.
- ☆ After 6 days, place the pro-trays with germinated seeds individually on the raised beds inside the shade net.
- ☆ Water with rose-can every day and drench with 19: 19: 19 @ 0.5 per cent (5g/l) at 18 days after sowing.
- ☆ Field preparation: Thoroughly prepare the field with the addition of FYM @ 25 t/ha and form ridges and furrows at a spacing of 60 cm.
- ☆ Apply 2 kg/ha of *Azospirillum* and 2 kg/ha of Phosphobacteria by mixing with 50 kg of FYM.
- ☆ Irrigate the furrows and transplant 30-35 days old seedlings at 60 cm apart on the ridges.
- ☆ Spacing: Varieties: 60 x 60 cm Hybrids: 90 x 60 cm
- ☆ Mulching: Mulch with black LDPE sheets of 25-microns thickness and bury both ends into soil upto the depth of 10 cm.
- ☆ **Weed control**: Apply Pendimethalin 1.0 kg a.i./ha or Fluchloralin 1.0 kg a.i/ha as pre-emergence herbicide,

- **Irrigation:** After establishment of seedlings, irrigate at weekly intervals.
- Apply FYM @ 25 t/ha as basal dose before last ploughing.
- Apply 2 kg/ha of *Azospirillum* and 2 kg/ha Phosphobacteria by mixing with 50 kg of FYM.
- Before planting, wet the beds using drip system for 8-12 hrs. Planting to be done at a spacing of 90x60x75 cm in the paired row system, using ropes marked at 75 cm spacing.
- Spray Pendimethalin 1.0 kg a.i./ha or Fluchloralin 1.0 kg a.i/ha as pre-emergence herbicide at 3rd day after planting.
- Gap filling to be done at 7th day after transplanting.
- Manuring Apply 2 kg each of Azospirillum and Phosphobacteria in the main-field at planting.
- Top dressing: 50 kg N/ha on 30th day of planting or during earthing up.
- Hybrids: Basal dose FYM 25 t/ha, NPK 100: 150: 100 kg/ha.
- Top dressing- 100 kg N/ha on 30th day of planting or during earthing up.
- Growth regulators: Spray 2 ppm (1 ml in 500 lit.) **Triacontanol** plus Sodium Borate or Borax 35 mg/l of water 15 days after transplanting and at the time of full bloom to increase the yield.
- **Epilachna beetle:** Insecticide Dose Azadirachtin 0.03 per cent WSP (300 ppm) 5.0 g/lit. Quinalphos 20 per cent AF 1.7 ml/lit.
- Whitefly: Spray Neem oil 3 per cent plus Teepol 1 ml/lit.
- Insecticide dose Diafenthiuron 50 per cent WP 8.0 g/10 lit. Phosphamidon 40 per cent SL 1.5 ml/lit. Thiamethoxam 25 per cent WG 4.0 g/10 lit.
- Shoot and fruit borer: Spray Neem Seed Kernel Extract 5 per cent
- Aphid: Apply phorate 10 per cent G @ 15 kg/ha Phosphamidon 40 per cent SL 1.5 ml/lit. Thiometon 25 per cent EC 1.0 ml/lit.
- Red Spider mite: Apply phorate 10 per cent G @15 kg/ha Insecticide dose Dicofol 18.5 per cent SC 2.0 ml/lit.
- Damping off: Treat the seeds with *Trichoderma viride* 4 g/kg or *Pseudomonas fluorescens* @ 10g/kg of seed 24 hours before sowing
- Drench with Copper Oxychloride at 2.5 g/lit.at 4lit./qm.
- **Leaf Spot**: Spray Mancozeb @ 2 g/lit.
- Little Leaf: Remove the affected plants in the early stages and spray Dimethoate 30 EC @ 1.0 ml/lit. to control the vector.
- Market information: Smooth and tender. Colour: light purple or dark purple, green, purple coloured with white stripes. Size: 25-30 cm long, oblong or round.

- IPM Package for Brinjal: Seed treatment with *Trichoderma viride* (4g/kg) and *Pseudomonas* @ 10 g/kg of seed Nursery + seedling deep treatment with *Pseudomonas* @ 10 g/lit. of water.
- Soil application with Neemcake @250 kg/ha

Chillies: *Capsicum annuum* L.; Solanaceae

- Chromosome no.- 2n=2x=24
- Origin- Mexico
- Primary centre of origin Mexico, Secondry Guatemala
- Non-climactric fruits, type of fruit-Berry
- Highest pungent chilli in world: tezpur chilli *C. Fruitscence* var. *nagahari.*
- Pungency rate = 855000-1000000 SHU (**scovile heat unit)**
- Highest vit-C-Red ripe chilli fruit then mature green fruit
- *C.annum and C.fruitscence* have white flower, while *C. Pubescence* have purple flower.
- China is major capsicum (bell paper) producing country
- India is major producer, consumer and exporter of chilli in the world
- June-Oct. is major chilli growing period in south India
- 7 gm good quality seed provide 100 plants
- Dry chilli contains about 6 per cent stalk 40 per cent pericarp 54 per cent seeds
- Chilli has bi-directional root system
- Highly pungent part of chilli **placenta**
- Vability of seed- 2 year
- Capsicum among green chilli - 210mg/100gm
- 36 per cent found colour mettel (*Capsanthin* colouring agents)
- Paperika-*Capsicum frutescence*
- Arkaabir suitable for colour extraction
- Ideal temp. 26-28 °C day time, night time 16-18 °C
- Chilli flower drop when temp. more 35 °C
- Emasculation is done in afternoon and pollination should be done in the next morning (forenoon 8-10am)
- Taxonomic character of five cultivated spp. Green leaf (1986)
- Extensive cross ability studies done by Smith and Heiser (1957)
- Green chilli to dry chilli ratio 10: 1, male and female ratio (2: 1)
- India shares in total export of chilli in the world is 4 per cent
- Fruit drop – **NAA triconital** is used to control fruit crop.

- HP is leading state in supplying capsicum during the off season
- 15 irrigations are required in capsicum
- Flower opens at 5 AM
- Chilli is often cross but cross pollinated is reported upto 63 per cent.
- Male sterility -CGMS
- Precursor of capsicum vanillylamine (isocarpic acid).
- 100kg of ripe fruit give 20-40kg of dry chilli.
- Pollination agent- Honey Bee, ants, thrips
- Chilli ripe fruit drying temp. 54.4 ^{0}C (2-3 days)
- Above 40 ^{0}C fruit drops
- Yield of green chilli is 7.5-10 ton/ha in irrigated condition
- 1.5-2.5 ton/ha in rainfed condition, dry chilli 2-2.5 ton/ha
- **Yellow anther chilli used in marker gene**
- Highest oleoresin content
- Ornamental peppers are also called Christmas pepper
- CMS in 1st discovered by Peterson in *C.annum* (PI164835)
- G- series lam selction research station (A.P)
- L series Pant nagar
- K series kovilpatti (TN)
- Isolation distance for foundation seed 400m and for Certified seed 200m
- Chilli breeder Dr. DP Singh
- Birds' chilli has highest capsicum content of 0.75 per cent.
- Capsicum var.
- **Introduction-** California Wonder, Yolo Wonder, Golden Wounder, Chines Giant, **Sweet Banana**, World Beater, **Bullnose,** Rabi King, **King of North**
- **Selection-**Arka Mohini/Arka Gaurav/Arka Basant
- Haihua-3- developed by- **Anther culture technique**
- **Public sectore hybrid-**
- Pusa Deepti, Green Gold
- Private sector var.- Early Bounty/Bharat/India/Hira/Lobia Lario
- **Chilli var.-**
- Selection- Kanpur yellow//Arka Lohit
- Bhagyalaxmi, Sindhure

- ☆ Hybrid- G-5/Andhra Jyoti- cherry type fruit
- ☆ Pusa wala- NP-46-A×Puri Red
- ☆ Pant C1- NP-46-A×Kandhari
- ☆ **Mutant- MDU1**
- ☆ **Private**- champion, **Tejaswini**/Gayatri Agni/Delhi hot
- ☆ NP46-1- Pureline selection- Tolerant to virus
- ☆ G1,G2,G3,G4- Pureline selection
- ☆ K-1 - Tolerant to thrips
- ☆ Sindhure- Mid pungent var.
- ☆ **Hybrid-**
- ☆ Pusa Sadabahar Pusa Jwala×IC31339- virus rest
- ☆ Punjab Lal- Perennial×Long Red
- ☆ Bhaskar- G4×Yellow anther mutant
- ☆ 1st private sectore hybrid- Bharat released in IAHS- company
- ☆ ArkaHarita- F1 hybird, Arka Meghana- F1 hybrid
- ☆ PusaSadabahar- Multipurpose var.
- ☆ JCA-154- Pickling var.
- ☆ 1st -F1 hybrid public sectore- PusaMeghdoot- *capsicum var.*
- ☆ NAA-500 ppm- fruit/flowering
- ☆ Fruit treated with- GA_3@ 200ppm or kinetin @ 100ppm
- ☆ Dropping fruit var.- Yolo Wonder
- ☆ **Yellow anther chilli used in marker gene**
- ☆ Highest oleoresine content
- ☆ Leaf curl resist var.- Pusa Jwala/Sadabahr/Pant C1
- ☆ Multiple disease resistant- T.M.V/Wilt/Dieback resistant var.- Punjab Lal
- ☆ Bacterial wilt resistant var.- Punjab Surkha/Utkal Rashmi/Ark Gaurav
- ☆ Jwalamukhi- **Suitable for HDP**
- ☆ High yield superior qualityvar.- Pusa Sadabahar
- ☆ Ornamental peppers are also called- christum pepper
- ☆ CMS in pepper- 1st discovered by- Peterson in *C.annum* PI164835
- ☆ Botanical name of bird pepper- *Capsicum frutescence*- Highly pungent
- ☆ Cherry pepper- *C. annum var.- cerasiforme*
- ☆ Cone pepper- *C.annum var.fasciculatum*

- ☆ Blue anther found- flower – *C.annum*
- ☆ Purple colour found flower- *C. annum pubescence/frutescence*
- ☆ Isolation distance- 400m for foundation and for Certified 200m
- ☆ Spacing-45×45 cm- solan condition
- ☆ 50×30 cm- banglore
- ☆ 80×90×30 hybrid
- ☆ Root pruning during cultivation can also caused blossom end rot
- ☆ Byadgi chilli has highest- pascicin content of 0.75 per cent
- ☆ Storage-
- ☆ Chilli-green- temp-7.2 ^{0}C and RH-85-90 per cent
- ☆ Chilli red-7.6-7.2 ^{0}C and RH 90-95 per cent
- ☆ Tolerant to soil acidity- **Moderate** pH 6.8-5.5
- ☆ Tolerant to soil salinity- moderate
- ☆ Water requirement- moderate
- ☆ Root depth- 90-120cm
- ☆ Ancestor of bell pepper- *Capsicum annum var.aviculare*
- ☆ Chilli found in India- 16th century
- ☆ Test weight- pungent type- 3.5gm, 5gm of sweet type
- ☆ Day natural plant- according to photoperiod
- ☆ Vit.C increase when temp. high by heating
- ☆ Seed yield- 100-200kg/ha
- ☆ Small fruit pungent type- 25-100g seed/kg fruit
- ☆ Sweet pepper yield- 5.50g seed/kg fruit
- ☆ **Disease-** anthracnose or dieback- *Colletotrichum capsici*
- ☆ Leaf curl virus transmitted by white fly
- ☆ Mosaic virus trans. by- Aphid
- ☆ **Frog eye leaf spot**- *Cercospora capsica*, most common nursery disease
- ☆ Thrips- *soirtothrips dorsalis*
- ☆ Physiological disorder- Blossom end rot- excess N_2+ water stress
- ☆ Dwarf ornamental var.- Alppers/marbles/riot
- ☆ Fresh green segment- Garima, Sitar, Japanese heat
- ☆ Dual purpose- HPH-12, HPH-17,HPH -4884
- ☆ Paprika cultivars- Arka Abir, KT-19, Bydogi, Warragal

- ☆ Pusa Jwala- NP-46-A× Pusa Red
- ☆ Jawahar 218- Kalikeeth×Pusa Jwala
- ☆ Punjab Lal-Perennial×Long Green
- ☆ **Soil:** Well-drained loamy soil rich in organic matter with pH range 6.5-7.5.
- ☆ **Season of sowing**: January – February, June – July and September - October
- ☆ **Nursery area:** 100 sq.m/ha.
- ☆ **Seed treatment**: Treat the seeds with Trichoderma viride @ 4 g/kg or *Pseudomonas fluorescens* @ 10g/kg and sow in lines spaced at 10 cm in raised nursery beds and cover with sand. Watering with rose can has to be done daily.
- ☆ Drench the nursery with copper oxychloride @ 2.5g/l of water at 15 days interval against damping off disease.
- ☆ Apply Carbofuran 3g @10 g/sq.m. at sowing.
- ☆ Protected nursery: Prepare the nursery area of 3 cents with slanting slope of 2 per cent for the seedling production to cover 1 ha.
- ☆ Water with rose can everyday upto seed germination. Drench with 19: 19: 19 @ 0.5 per cent (5g/l) at 18 days after sowing.
- ☆ **Field preparation:** Thoroughly prepare the field with the addition of FYM @ 25 t/ha and form ridges and furrows at a spacing of 60 cm.
- ☆ Apply 2 kg/ha of *Azospirillum* and 2 kg/ha of Phosphobacteria by mixing with 20 kg of FYM.
- ☆ Irrigate the furrows and transplant 40-45 days old seedlings, with the ball of earth on the ridges.
- ☆ Spacing: Varieties: 60 x 45 cm Hybrids: 75 x 60 cm
- ☆ Weed control: Apply Pendimethalin 1.0 kg a.i./ha or Fluchloralin 1.0 kg a.i./ha as pre-emergence herbicide followed by hand weeding once 30 days after planting.
- ☆ Irrigation: Irrigate at weekly intervals.
- ☆ Apply 2 kg/ha of *Azospirillum* and 2 kg/ha Phospho bacteria by mixing with 20 kg of FYM.
- ☆ Apply 75 per cent total recommended dose of superphosphate i.e. 375kg/ha as basal dose.
- ☆ Install the drip irrigation with main and sub main pipes and place lateral tubes at an interval of 1.5 m.

- ☆ Form raised beds of 120 cm width at an interval of 30 cm and place the laterals at the centre of the each bed.
- ☆ Before planting wet the beds using drip system for 8-12 hrs.
- ☆ Planting to be done at a spacing of 90x60x45 cm in the paired row system, using ropes marked at 60 cm spacing.
- ☆ Spray Pendimethalin 1.0 kg a.i./ha or Fluchloralin 1.0 kg a.i/ha as pre-emergence herbicide at 3rd day after planting.
- ☆ Gap filling to be done at 7th day after transplanting.
- ☆ Manuring: Basal dose - FYM 25 t/ha, NPK 30: 60: 30 kg/ha.
- ☆ Potassium as K_2SO_4 for quality improvement. Application of potassium in the form of potassium sulphate will increase quality of chilli.
- ☆ Top dressing: 30 kg N/ha in equal splits on 30, 60 and 90 days after planting.
- ☆ Hybrids Basal dose: FYM 30 t/ha, NPK 30: 80: 80 kg/ha.
- ☆ Top dressing: 30 kg N/ha in equal splits on 30, 60 and 90 days after planting.
- ☆ Fertigation schedule - Chilli F1 hybrid Recommended Dose: 120: 80: 80 kg/ha
- ☆ Effect of Endo root soluble and Mycorrhizae on chilli Apply 250g of Endo Roots Soluble in two splits doses at 15 DAT and 45 DAT along with 100 per cent N and K and 50 per cent P for higher yield and saving of Phosphorous.
- ☆ Apply 250g of Mycorrhiza in two splits at transplanting and 30 DAT along with 100 per cent N and K and 50 per cent P for higher yield and saving of Phosphorous.
- ☆ **Growth regulators:** Spray *Triacontanol* @ 1.25 ml/l on 20, 40, 60 and 80th day of planting. Spray NAA 10 ppm (10 mg/l of water) on 60 and 90 days after planting to increase fruit set.
- ☆ Micronutrient spray: Foliar spray of $ZnSO_4$ @ 0.5 per cent thrice at 10 days interval from 40 days after planting.
- ☆ Spray 19: 19: 19 + Mn @ 1 per cent at 60 days after planting.
- ☆ Weed control: Spray Fluchloralin @ 1 lit./ha or Pendimethalin @ 1 kg a.i/ha.
- ☆ Raise intercrop of onion in paired row system to get additional income.
- ☆ **Fruit borer**- Set up pheromone traps for ***Helicoverpa armigera/ Spodoptera litura*** @ 12 no./ha.Spray *Bacillus thuringiensis* @ 2 g/lit.
- ☆ **Thrips:** - Apply carbofuran 3 per cent 2x2 @ 33 kg/ha or phorate 10 per cent @ 2x2 10 kg/ha.

- **Aphids**: Treat seeds with imidacloprid 70 per cent WS @12 g/kg of seed. Apply phorate 10 per cent @ 10 kg/ha
- Yellow Muranai mite: Apply phorate 10 per cent @ 10 kg/ha Insecticide dose Buprofezin 25 per cent SC 8.0 ml/10 lit.Diafenthiuron 50 per cent WP 8.0 g/10 lit.
- **Aphids and Thrips: Spray neem oil 1 per cent or neem cake extract 5 per cent to control aphids and thrips.**
- Damping off: Treat the seeds with *Trichoderma viride* 4 g/kg or *Pseudomonas fluorescens* 10 g/kg of seed 24 hours before sowing.
- Leaf spot: Spray Mancozeb @ 2 g/lit. or Copper oxychloride @ 2.5 g/lit.
- **Powdery mildew:** Spray Wettable sulphur @ 3 g/lit. or Carbendazim @ 1 g/lit., 3 sprays at 15 days interval from the first appearance of symptom.
- **Die-back and fruit rot**: Spray Mancozeb 2 g/lit. or Copper oxychloride @ 2.5 g/lit. thrice at 15 days interval starting from noticing the die-back symptoms.
- **Chilli mosaic**: Raise 2 rows of maize or sorghum for every 5 rows of chilli crop against the wind direction.
- Yield of varieties: 2 - 3 t/ha of dry pods or 10 - 15 t/ha of green chilies.
- Dry Chilli: Well dried -12 per cent moisture,Big size Bright red colour,without white chillies, Uniform size Length - 1.5-2 inches, Green chilli Pungency

Capsicum (Sweet Pepper/Bell Pepper): *Capsicum annuum* var. *grossum*; Solanaceae

- **Varieties**: Arka Basant, Arka Gaurav, Arka Mohini, Green Gold, Bharat.
- **Soil:** Well drained loamy soil rich in organic matter. It can be grown on any fertile well-drained soil suitable for chillies cultivation with pH 6.5 - 7.0.
- Season of sowing: September - February and June - July.
- **Seed rate**: Varieties: 1.25 kg/ha,Hybrids: 200 g/ha
- Seed rate: 500 g/ha.
- Spacing: 60 x 45 cm
- Nursery: 3 cents/ha Treat the seeds with Carbendazim @ 2 g/kg of seed and sow in lines across the bed at a spacing of 2.5 cm and then cover with top soil and then paddy straw. Watering with rose can has to be done daily.
- On 20th day of sowing, 300g of carbofuran 3g granules have to be applied in between the seedling lines across the bed, the soil has to be stirred and then the beds are irrigated.
- Cover the nursery area with 50 per cent shade net and cover the sides using 40/50 mesh insect proof nylon net.
- Form raised beds of 1m width and convenient length and place.
- 238 protrays (98 cells) are required for the production of 23334 seedlings, which are required for one hectare adopting a spacing of 90x60x60 cm in a paired row system.
- Water with rose-can everyday and drench with 19: 19: 19 @ 0.5 per cent (5g/l) at 18 days after sowing.
- Preparation of field Plough the field to a fine tilth. Form ridges and furrows 45 or 60 cm apart.
- Transplanting: 40-45 days old seedlings at 30 cm spacing.
- Irrigation: Irrigation at weekly or 10 days interval.
- Apply FYM @ 25 ton/ha as basal before last ploughing.
- Spray Pendimethalin 1.0kg a.i./ha or Fluchloralin 1.0kg a.i/ha as pre-emergence herbicide at 3rd day after planting.
- Gap filling to be done at 7th day after transplanting.
- Application of fertilizers - Apply FYM 25 t/ha and NPK 40: 60: 30 kg/ha as basal and 40 kg N/ha each on 30, 60 and 90 days of planting as top dressing.

- ☆ On 30th day, hoeing and weeding has to be done once and the plants are earthed up.
- ☆ Growth regulator: Spray 1.25 ppm Triacontanol (12.5 mg/10 l of water) on 20th, 40th, 60th and 80th day after transplanting. Spray NAA 10 ppm (10 mg/l of water) on 60 and 90 days after planting.
- ☆ Powdery mildew: Spray wettable sulphur @ 2g/l
- ☆ Dieback and fruit rot: Spray Mancozeb @ 2g/l
- ☆ Anthracnose: Spray Mancozeb @ 2 g/l.
- ☆ Fruit rot: Spray Copper oxychloride @ 2.5 g/l.
- ☆ Harvest and yield: Harvest fully matured green fruits before ripening.
- ☆ Yield: 15 tonnes/ha in 150-160 days.

Potato: *Solanum tuberosum* L.; Solanaceae

- ☆ Origin: South India 2n=48
- ☆ Flower- Harmaphordite
- ☆ Proginator of potato ***-Solanum stenotonum***
- ☆ Edible part- Modified under ground stem
- ☆ Long day plant.
- ☆ 1st mention of potato in India by Asaf khan/sir Thomas roe in 1615
- ☆ **CPRI was established in 1949**
- ☆ Potato breeding station Khufri, Shimla was established in 1935 under IARI
- ☆ Inflorescence –Cymose
- ☆ Potato introduced in 17 Century by Portuguese from Europe
- ☆ 1st CPRI director **S. Ramanujan**
- ☆ Temp. for tuber germination 15-25 ^{0}C
- ☆ Vegetative growth 20 ^{0}C low temp.
- ☆ Tuberization and tuber development high light intensity and short day
- ☆ Day temp. 20 ^{0}C and 4 ^{0}C night temp.
- ☆ **Suberization of potato 25 ^{0}C**
- ☆ Sprouting of potato is best 22-34^{0}c
- ☆ 30^{0}C completely stop the tuber formation
- ☆ Hip and pit method for potato storage
- ☆ Alkaline soil favours the development of scab disease
- ☆ Well pulverize soil facilitate batter tuberization
- ☆ Parent of solanum tuberosum **(*S. stenotonum x S. sparipillium*)**
- ☆ Seed treatment: deep the seed 0.2 per cent solution Areton and Aglol 6
- ☆ Dithane M 45 for 20-30min.
- ☆ Treatment of whole seeds with ethylene chlorohydrine at 35-72 hrs.
- ☆ Finally treat GA_3 1 per cent for 10 min.
- ☆ Water requirement 350- 550 mm
- ☆ Bitter taste due to presence of solanin
- ☆ Only 45 per cent potato accommodate in cold storage
- ☆ **SPT was developed by Puskarnath 1965**

- ☆ SPT space 50x25cm and diameter 4-6cm
- ☆ Dehaulming removal of the top portion of potato in the seeds
- ☆ Dehaulming should be done 10-12 days before harvesting
- ☆ Ratio of male and female of TPS is 1: 4
- ☆ TPS required 100-120gm/ha for nursery raised bed
- ☆ TPS 1gm seed -1500-2000 yield 250- 300q/ha 30 per cent large size of tuber used as table purpose
- ☆ They are ready (4-5 leaf) for transplanting
- ☆ Potatos are shallow rooted
- ☆ **International potato year- 2008**
- ☆ Potato is C_3 plant
- ☆ Respiration rate is very slow
- ☆ **Aphid infection 20 aphid/100 leaves**
- ☆ India rank 3rd in productivity of potato
- ☆ Collection of potato tuber- Hyping
- ☆ Optimum temp for potato seed germination is 18- 20 °C
- ☆ Starch content 60-80 per cent, CHO 22.6 g/100g
- ☆ Potato is self pollinated crops
- ☆ Potato is rich source of vit- C
- ☆ Potasium increases the no. of large size tuber
- ☆ Phosphorus increases the no. of medium size tuber
- ☆ **AICRP on Potato 1971**
- ☆ **International potato centre Lima in peruin 1971**
- ☆ **Solanin present in potato = 5mg/100gm**
- ☆ 75-80 per cent total dry matter contain starch
- ☆ Protein contain 1.6 per cent sprouting can be checked by MH 0.3 per cent
- ☆ Photo insensitive var. Kufri Alankar
- ☆ Isolation distance 10m for foundation and 5m for certified seed
- ☆ Isolation distance for TPS- 50m
- ☆ Black heart is resistant by sub- Oxidation
- ☆ Pollen can store in refrigerator 1-2 week and freezer 6-12 month
- ☆ **Potato can be grown all state except Kerala**
- ☆ Potato tuber dormancy 8-10 weeks

- ☆ Hills account for only 5 per cent of area under potato
- ☆ Ridge and furrow method most popular method of potato planting
- ☆ N_2 is most important nutrient for potato tuber.
- ☆ Overy type (hypogyuous, superior)
- ☆ Earthing up of potato is done 40 days after sowing
- ☆ 27.4 per cent total veg. production 22.9 per cent potato largest area cultivation
- ☆ ***Solanum phureza*** is useful for high protein spp.
- ☆ ***Solanum demisom*** is original source of late blight
- ☆ Pentaploid spp. ***Solanum curtilobum*** 2n= 60
- ☆ ***Solanum maglia*** is drought resistant
- ☆ ***Solanum verni*** high starch containt spp.
- ☆ **Diploid spp. *Solanum stenotonum, phureja*, *ajanhuiri*** (75 per cent majority)
- ☆ **Triploid spp. *Solanum chaucha, Solanum juzepczukii***
- ☆ **Tetraploid spp. *S. tubrosum* var. *tubrosum,*** *S. audigena*, *S. acaulie*
- ☆ **Pentaploid spp. *Solanum curtilobum***
- ☆ Kufri Kuber is resistant to black scurf
- ☆ **Kufri Badshah is multiple resistant to frost and late blight**
- ☆ Kufri Sheetman is resistant to frost
- ☆ Kufri Chipsona 1,2 and Kufri Anand are suitable to chips
- ☆ Cyst nematode resist. var. –Kufri Suwarna, Aanamalai
- ☆ Potato is dicot plants
- ☆ True potato hybrid seed cultivar. HPS-1/13, HPS-11/13,HPS-24/11
- ☆ Resist. to late blight kufri Jyoti, Moti/Kundan/Jeevan/Naveen
- ☆ Kufri deva more tolerant frost and rotting
- ☆ Early blight – Sindhuri, Kufri Pukhraj
- ☆ Kufri Kuber – Resistant to black scurf
- ☆ Kufri Lalima is a now replacing Kufri Sindhuri
- ☆ Early var. (80-90 days)
- ☆ Kufri Chandramukhi (1968) and K Alankar early blight – early maturity var.
- ☆ Kufri Jawahar (1996) Kufri Lavkar 1972 Kufri Kuber, Kufri Sheetman, K Jyoti/K. Surya, K Ashoka

- Mid var. Kufri Bahar/Kufri Satlaj/Kufri Anand/Kufri Jyoti (1968)/Kufri Arun/Kufri Kanchan,
- K. Pukhraj, K. Puskar, Kufri Muthu 1971
- Late var.— Kufri Deva/Kufri Megha
- Clonal selection — Kufri Red/Kufri Safed
- Multi resist. var. Kufri Badshash
- Hybrid var.-
- Kufri Chandramukhi (S.-448 X Kufri Kuber)
- Kufri Badshah (Kufri Jyoti x Kufri Alankar)
- Kufri Sinduri (Kufri Red x K Kanchan)
- Kufri Jawahar (Kufri Neelmani x K Jyoti)
- Kufri Seetman (Phulwa x Criags difence)
- Kufri Satlaj (Kufri Bahar x Kufri Alankar)
- Berry drop is common Problem in potato hybridization – 2-4, D solution for control
- **Disease**
- Late blight of potato (phytopthora infestence)
- Wort disease- (***Synchytrium indobioticum***), resistant var. Kufri Jyoti/ Kufri Kanchan/Sherpa
- Black scurf –***Rhizoctoni solani***, ressitant– var.– Kufri Kuber
- Black wilt- ***Ralastonia solanacearum***
- Soft rot – ***Erwinia carotovera***
- Potato tuber moth – ***Phythorimaia operculella***, 30-70 per cent damage tuber storage
- Cyst nematode- ***Globodera ssp.*** Major pest in decconplateue
- Greening: exposure to sunlight
- Internal brown spot- Moisture deficiency
- Black heart – Poor ventilation
- Hollow heart - Excessive N2
- Chilling injury- Low temp. (^{0}C)
- Freezing injury- (Low temp. -1 to ^{0}C)
- Cultivated in regions receiving a rainfall of 1200 - 2000 mm per annum.
- Potato is mostly grown as a rainfed crop.
- Season and planting-Hills Summer: March – April

- Autumn- August – September
- Irrigated - January – February
- Plains: October – November. Use disease free, well sprouted seeds weighing 40 – 50 g.
- Use Carbon disulphide 30 g/100 kg of seeds for breaking the dormancy and inducing sprouting of seeds.
- Plant the tubers 20 cm apart.
- Seed rate: 3000 – 3500 kg/ha.
- Preparation of field: Prepare the land to fine tilth.
- In hills provide an inward slope of 1.40 in the terraces.
- Provide drainage channel along the inner edge of the terrace.
- Form ridges and furrows with a spacing of 45 cm between ridges.
- Irrigation: Irrigate the crop 10 days after planting.
- Application of fertilizers: Apply 15 t of FYM/ha, and 2 kg each of Azospirillum and Phosphobacteria as basal and 120 kg N, 240 kg P and 120 kg K/ha in two splits
- Spacing: 60 x 30 x 20cm in paired row system
- Spray of Gramaxone @ 2.5 l/ha in 500 lit. of water as post - emergence.
- **Cut worms:** Fork the soil during summer months to expose the larvae and pupae for avian predators. Set up light trap between 7-10 pm to attract the adults.
- Set up sprinkler system during day time to expose the hidden larvae for predation by birds. Apply NSKE 5 per cent or neem oil @ 2 lit./acre using 500 lit.of spray fluid.
- **Potato tuber moth:** Select healthy tubers and plant at 15 cm depth to avoid oviposition.Set up pheromone traps @ 5/acre. Spray quinalphos @ 1 lit./ha at ETL of 5 per cent leaf damage
- **Aphids:** Apply carbofuran 3 per cent g @ 17 kg/ha or Phorate 10 per cent @ 10 kg/ha or drench thiamethoxam 25 per cent WG @ 200 g in 500 lit. of water per ha
- **Late blight:** Carefully select seed tubers. Remove ground creepers which serve as a source of infection.
- Spray Mancozeb 2 g/l or Chlorothalonil 2 g/l on 45, 60 and 75 days after planting. Grow late blight resistant Varieties like Kufri Jyothi, Kufri Malar and Kufri Thangam.
- **Brown rot:** Select disease free seeds. Give proper drainage facilities. Remove and destroy the affected plants.

- **Early blight:** Spray Mancozeb 2 g/l or Chlorothalonil 2 g/l at 45, 60 and 75 days after planting.
- **Virus diseases:** Use virus free potato seeds. Rogue the virus affected plants regularly. Control the aphid vectors by spraying dimethoate 30 EC or methyl demeton 25 EC @ 2 ml/lit.
- Yield: 15 – 20 t/ha in a duration of 120 days.

Chapter 2

Cole Crops

Cauliflower: *Brassica oleracea* var. *botrytis*; Brassicaceae

- Chromosome number 2n=18
- **Cauliflower is a thermosensitive crop**
- Edible part of cauliflower is curd
- Cauliflower was introduced to India 1822 Dr. Jenson
- Ancestore of cauliflower is ***Brassica cretica***
- Type of infloresecence- Racemose
- **Protogyous in nature**
- Early cauliflower and sprouting broccoli- Annual nature
- Late type- Snow ball have self blanched habits
- Multi- allelic single gene
- Optimum temperature for curd initiation 17-20 ^{0}C
- Common herbicides use in cabbage- Basalin 3.3 lit./ha
- Storage temp 0 ^{0}C and 90-95 per cent RH
- Sccoping is sponical operation done in cauliflower- west Bengal
- Seed rate 500-600gm early Varieties
- Mid and late 350-400 gm/ha
- Self banched variety of cauliflower – Pusa Dipali
- Self blanched and off season – Pusa Himjyoti, Hisar 1

- ☆ Tolerant to curd and inflorescence blight- Pusa Synthetic
- ☆ Highly resistant to black rot and free from ricyness- Pusa Ketki
- ☆ Cauliflower bear harmaphordite flower
- ☆ Resistant to black rot – Pusa Snowball K1
- ☆ Tropical cauliflower variety- Arka Kanti
- ☆ Synthetic variety- Pant Gobhi-3
- ☆ Extra early-end of Feb, temp. 20-30 ^{0}C, harvest-May-June
- ☆ Ex-Early: Kumari, Pusa Meghana, Pusa Kartik, Sankar
- ☆ Extra early 1- End of may temp.- 22-27 ^{0}C
- ☆ Ex-Pant Gobhi-3, Pusa Early Synthetic, Pusa Meghna, Arka Kranti
- ☆ Availability- end August-Oct.
- ☆ Extra early 2- temp. 20-25 ^{0}C end of Oct-Nov
- ☆ Sowing- June Pusa Deeplai, Pusa Ketki- Jan.
- ☆ Pusa Synthetic, Pusa Himjyoti, Pusa Subra, Punjab Gaint-35
- ☆ Late sowing- Sept-Nov 10-16 ^{0}C, available- Jan-March
- ☆ Pusa Snow Ball 1, K1, 16, Ooty-1
- ☆ F1 hybrid- Pusa hybrid 2 (CC-35×18-19)
- ☆ Pusa Kartik Sankar-(CC-14×41-5) - Resistant to downey mildew
- ☆ Pusa Snow Ball 1= EC12012×EC12013
- ☆ Pusa Snow Ball K1= Pusa Snow Ball×Dontch
- ☆ Snowball-16- Introduced from Holland
- ☆ 6- inbred line used in Pusa Synthetic variety
- ☆ Palam Uphar
- ☆ Seed yield of cauliflower 400kg/ha
- ☆ 1 gm seed contain 300-350 seed
- ☆ 1-2 per cent area used for curd initiation
- ☆ Isolation distance- foundation 1600m, cerified-1000m
- ☆ Pusa hybrid 2- 1st public sector hybrid
- ☆ Pusa Himjyoti- sowing time April- July in hill
- ☆ Arka Vimal and Arka Spurthi- Seed set under tropical land
- ☆ Snow ball seed set only hill
- ☆ 5-8 irrigation during growth
- ☆ Molybdenum and boron very important micro nutrient

- Sclerotia rot – Resistant to Pusa Snowball K-25
- Pusa Synthetic- 6-8 inbreed line use
- Pant Gobhi-3 of inbred line
- Pant Gobhi-4 recurrent selection
- Test weight -2.8gm
- Cauliflower is a long day plant
- Black rot resistant variety- Pant Subra, Pusa Snowball K1
- Dania is a variety of cauliflower
- Late type snowball variety- C_6- INDEX
- 2 mutant gene BOAP-1-a and Bola-1-1
- Buttoning is due to nitrogen deficiency
- Chlorosis due to Mg deficiency
- Browning of curd- Boron deficiency in alkaline soil – borax 10-12kg/ha
- Black rot- ***Xanthomonas compestris***- Most important disease
- Black rot- Seed borne bacterial disease of cauliflower
- Good source – glicosinolate- 40-8-mg/100gm
- Sinigrin-Glucobrassicin (anti-carcinogenic)
- Snowball type curd initiation 10-16 ^{0}C
- Indian or tropical type 20 ^{0}C or above.
- **Nursery:** 100 sq.m nursery area/ha.
- Apply FYM at 300 kg and 10 kg of No.5 mixture (9: 9: 9) along with 50 g of sodium molybdate and 100 g of borax.
- Sow the seeds at 10 cm between rows in raised seed beds after drenching with copper oxychloride (2.5 g/l).
- Transplant 30 to 40 days old seedlings at a spacing of 45 cm. Avoid land infected with 'club root disease'.
- Sow the seeds in raised beds and transplant 25 days (early Varieties), 45 days (late Varieties) old seedlings at 45 cm apart
- Seed treatment: 250 g of hybrid cauliflower seed is required for the production of seedlings for 1 ha. Treat the seeds in hot water @ 50 ^{0}C for 30 minutes.
- Sowing: Sow the seeds in protrays @ 1 seed per cell. Cover the seeds with cocopeat, keep the tray one over the other (8-10 No) and cover with polythene sheat for 5 days or till germination starts.

- ☆ Bring the soil to fine tilth. Pits should be taken at a spacing of 45 cm either way in hills. Form ridges and furrows at 60 cm in plains.
- ☆ Irrigation in Hills: Once in a week during January and February.
- ☆ Plains: Once in a week.
- ☆ Hills: Apply 30 t/ha of FYM and 90 kg N, 90 kg P and 90 kg K as basal dose and 45: 45: 45 kg NPK/ha after 45 days.
- ☆ Plains: Apply 15 t of FYM/ha and 50 kg N, 100 kg P and 50 kg K as basal and 50 kg N after 45 days.
- ☆ Fertigation: Fertigation requirement for F1 hybrid: 200: 125: 125 kg of NPK/ha. Apply once in every three days throughout the cropping period.
- ☆ Spacing: 60x45x45cm in paired row system.
- ☆ Gap filling after 20 days of planting to maintain the population and uniform growth.
- ☆ Hoeing and weeding on 30th and 45th day. Avoid deep intercultivation as it is a shallow rooted crop.
- ☆ Aphids: Install yellow sticky trap @ 12 Number/ha to monitor macropterous adults (winged adult).
- ☆ Apply phorate 10 per cent G @ 20 kg/ha or spray neem oil 3 per cent with 0.5ml teepol/lit. or spray Azadirachtin 5 per cent, Neem Extract Concentrate 5.0 ml/10 lit.
- ☆ Diamond backmoth: Grow mustard as intercrop at 20: 1 ratio. Install pheromone traps @ 12 No./ha. cartap hydrochloride @ 1 g/lit. or ***Bacillus thuringiensis*** @ 1g/lit.at primordial stage (ETL 2 larvae/plant)
- ☆ Club root: Biological control. Seed treatment with ***Pseudomonas fluorescens*** at 10 g/kg of seeds, followed by seedling dip @ 5g/l and soil application @ 2.5 kg/ha along with 50 kg FYM before planting.
- ☆ Leaf Spot: Spray Mancozeb at 2 g/l or Carbendazim 1 g/l.
- ☆ Leaf Blight: Spray Mancozeb @ 2.5 g/litre.
- ☆ **Blanching**: Blanching refers to covering of curds.
- ☆ A perfect curd is pure white. It is necessary to exclude sunlight to obtain this.
- ☆ The common practice is to bring the outer leaves up over the curd and tie them with a twine or rubber band.
- ☆ By using a different coloured twine each day.
- ☆ It is easy at the time of harvest to select those tied earlier.
- ☆ Browning or brown rot: This is caused by boron deficiency.
- ☆ It appears as water-soaked areas and later change into rusty brown.

- Spray one kg of borax in 500 litre of water 30 days after planting.
- **Whiptail**: This results from the deficiency of molybdenum. It is more pronounced in **acidic soil.**
- The leaf blades do not develop properly. In severe cases, only the midrib develops and it can be corrected by spraying 100 g of Sodium molybdate in 500 l of water 30 days after planting.
- **Buttoning:** The term buttoning is applied to the development of small curds or buttons.
- Deficiency of nitrogen and planting the early Varieties late may cause these symptoms. Avoid transplanting of aged seedlings.
- **Blindness:** Blind-cauliflower plants are those without terminal buds.
- The leaves are large, thick, leathery and dark green.
- It is due to the prevalence of low temperature when the plants are young or due to damage to the terminal bud during handling the plants or due to injury by pests.

Cabbage: *Brassica oleracea* var. *capitata*; Brassicaceae

- Ancestor- ***Brassica oleracea var. sylvestris***
- Chromosome number 2n=18
- It ranks second next to cauliflower in area and production
- Large grow in (southern eastern coastal asia)
- Area and production: it cover about 4 per cent area under vegetable crop in our country
- Heat tolerant and F_1 hybrid give more yield/unit area compare to traditional cultivar in area of mild winter
- 22.9 per cent hybrid cabbage grown.
- Cabbage is known to possess medicinal properties
- Cabbage have anti-cancer property
- It protects against bowel cancer due to the presence of ***indole-3 carbinol***
- Cabbage juice was use as remedy against
- Poisonous mushroom and has a gorgle against- hoarseness
- Cabbage leaf used for cover- ulcer and wound
- Red and white cabbage found in Germany 1150 AD, and 14th century in England
- Cabbage cultivated in India – Mughal period-1542
- Union Minister of India of Agriculture Govt. of India, started a station seed production technology of temperate veg. include cabbage H.P- **Kulloo 1949**
- **Type of fruit-Siliqua**
- Cabbage is a biennial crop
- *B.O.capitata- alba*-White cabbage
- *B.O.capitata – rubra*-Red cabbage
- *B.O.var.sabuda*- Savoy cabbage
- Cabbage is highly cross pollinated 73 per cent
- Stigma are active in flowering stage but remain inactive in bud stage
- Glycoprotein presents in stigma
- High temperature high humidity/high concentration of CO_2 is responsible for break down of SI

- ☆ Cytoplasmic genetic male sterility found in cabbage helpful for the production of F_1 hybrid
- ☆ Cabbage required specific low temperature for chilling 4-8 ^{0}C, means temp for 40-60 days so that plant grow to reproductive stage of flowering
- ☆ Maturity early type -55-70 day
- ☆ Late variety- 100-120 day
- ☆ Soil- cabbage does not grow well in highly acidic soil, pH 5.5-6.5
- ☆ C- index is means of the salt content
- ☆ Red cabbage is the normal limit of salt tolerant is (C4)
- ☆ Savoy cabbage it is C-6
- ☆ C-8 their may be reduction of 40 in yield
- ☆ Plant grown in saline soil are more susceptible to disease like black leg
- ☆ Climate-Japanese investigation 15-20 ^{0}C is the optimum temperature range for growth and head formation in cabbage
- ☆ Min temperature for growth cabbage in just above ^{0}C
- ☆ Cabbage may with stand temperature of 10-40 ^{0}C for few day
- ☆ Storage temperature 0-1.7 ^{0}C with 90-95 per cent RH
- ☆ Production of flower depend upon the temperature net on the day length
- ☆ Young plant can be grown at higher temperature
- ☆ Intensity of flower also depend on the age of the plant
- ☆ Optimum seed germination is obtained 12.8-15.6 ^{0}C
- ☆ Seed rate 375gm/ha- late
- ☆ Early : 500gm/ha
- ☆ Nursery bed : 109m/ha
- ☆ 37-50gm seed/ha
- ☆ Spacing- early 45×45cm, late 45×60
- ☆ Use of growth regulator
- ☆ Round oarmature earlier followed by cinicalvar.
- ☆ 50 ppm boric acid at flowering enhance the seed yield
- ☆ GA_3 100-1000ppm- result in better development of seed
- ☆ A spray of CCC or SADH 2500-5000ppm increase the low temperature resistant in cabbage
- ☆ Cabbage transplanting 4-6 true leaf stage
- ☆ Storage early variety storage 0-1.7 ^{0}C 92-95 per cent RH

- ☆ Late variety 12 weeks
- ☆ Low cost store is knowns as bricks and sand store
- ☆ P. Sauerkraut it is fermentated pickle prepared by white cabbage
- ☆ Soil pH 5.5-6.5
- ☆ 1st hybrid of cabbage 1950 Japan
- ☆ Use self incompatibility– **Nagaoka-1**
- ☆ Varieties-Golden Acre- selection from Coppenhagen Market
- ☆ Coppenhagen Market- Biggers variety popular in West Bengal
- ☆ Late maturing – Pusa Drum Head
- ☆ September introduced from Germany
- ☆ Conical type-Jercy Wake Field
- ☆ August- Early variety
- ☆ Glori of Eukhuizen/Red Acre
- ☆ Pride of India
- ☆ Selection- Pusa Ageti- 1st tropical var. developed in high temperature condition
- ☆ Hybrid- Pusa Drum Head- F1 hybrid from Japan
- ☆ **Black leg and dry rot resistant var.**
- ☆ Pusa Mukta- Black rot resistant var.- (EC24855×EC10109)
- ☆ Pusa Sambandh- Suitable for high density planting
- ☆ **Private sector variety-**
- ☆ Quento/Shree Ganesh Gol/Uttam/Bajarang
- ☆ Green Boy and Green Express- Hot wether hybrid 30-35°C
- ☆ Sudha, Stone Head, Green Chalanger, Green Cornet, Savarna, Vishses, Pragati
- ☆ Yamuna, Ganaga, Kavery- IAHS
- ☆ Isolation distance- Foundation 1600m, certified 1000m
- ☆ **Disease-**
- ☆ ***Pythium aphindarmatun***- Damping of major problems of nursery bed
- ☆ Caused by fungus
- ☆ Sclerotinia- ***Sclerotinia sclerotiorum***
- ☆ Commonly known as- white rot
- ☆ Serious in hilly area

- Club rot- ***Phasmadiphora brassicae***
- Black rot- ***Xanthomonas compestris*** – V shape chlorosis
- Soft root- ***Ervinia carotovora***
- **Insect pest-**
- Cabbage head borer
- Diamond black moth- ***Plutella xylostella***
- Tobacco caterpiller**- *Spodoptera litura***
- Hot water treatment 50 °C for 30 minute- seed borne disease- Black leg and black rot
- Rich in vitamin A and C
- Sinigrin toxic substance
- Yield-
- Early variety- 33-45tonnes/ha
- Hybrid -70-80tonnes/ha
- Nursery: 100 sq.m nursery area/ha. Apply FYM at 300 kg and 10 kg of No.5 mixture (9: 9: 9) along with 50 g of sodium molybdate and 100 g of borax.
- Sow the seeds in rows drawn at 10 cm spacing in raised seed beds after drenching it with copper oxychloride (2.5 g/l).
- Seedlings will be ready for transplanting in 40-45 days after sowing. Avoid land infected with 'club root disease'.
- Protray: The Protrays of 98 cells are ideal for cabbage seedling production.
- Around 600 protrays are required for the production of 28.333 seedlings required for one hectare at spacing of 60x45x45 cm in three row planting
- Seed treatment: 250 g of hybrid cabbage seed is required for the production of seedlings for 1 ha.
- Treat the seeds in hot water @ 50 ^{0}C for 30 minutes 25g of *Azospirillum* is required for the seed treatment of 250g cabbage seeds.
- Sowing: Sow the seeds in protrays @ 1 seed per cell. Cover the seeds with cocopeat and keep the tray one over the other (8-10Nos) and covered with polythene sheet for 5 days or till germination starts.
- The cabbage seedlings are ready for transplanting in 25 days
- Preparation of field: Bring the soil to a fine tilth. Pits should be taken up at a spacing of 40 cm either way in Hills.
- Ridges and furrows are formed at 45 cm apart in plains.
- Spacing Hills: 40x40 cm, Plains: 45x30 cm

- ☆ Irrigation: Provide continuous supply of moisture.
- ☆ Application of fertilizers: Hills: Apply 30 t/ha FYM, 90 kg N, 90 kg P and 90 kg K as basal and 45: 45: 45 kg NPK/ha 30 to 45 days after planting.
- ☆ Plains: Apply 20 t/ha of FYM. 50 kg N, 125 kg P and 25 kg K/ha along with 2 kg Azospirillum as basal and 50 kg N after one month of planting and earth up.
- ☆ Fertigation: Fertigation requirement for F1 hybrid: 200: 125: 150 kg of NPK/ha. Apply once in three days throughout the cropping period.
- ☆ Spacing - 60x45x45cm in paired row system
- ☆ Fertigation schedule: Recommended Dose: 200: 125: 150 kg/ha
- ☆ After cultivation: Deep hoeing should be avoided, as the Cabbage roots are surface feeders.
- ☆ Aphids: Install yellow sticky trap @12 no/ha to monitor "macropterous" adults (winged adult). Spray neem oil 3 per cent with 0.5 ml Teepol/lit.,
- ☆ **Diamond backmoth**: Grow mustard as intercrop at 20: 1 ratio to attract diamond back moths for oviposition.
- ☆ Release parasite Diadegmasemiclausum @ 50,000/ha, 60 days after planting.
- ☆ Spray NSKE 5 per cent after primordial stage ***Azadirachtin 0.***03 per cent WSP (300 ppm) 5.0 g/lit.
- ☆ Club root: Biological control Seed treatment with ***Pseudomonas fluorescens*** at 10 g/kg of seeds, followed by seedling dip @ 5g/l and soil application @ 2.5 kg/ha along with 50 kgFYM before planting.
- ☆ Leaf Blight: Spray Mancozeb @ 2.5 g/l.
- ☆ **Ring spot**: Spray Mancozeb 2 g/l or Carbendzim 1 g/l or Copper oxychloride 2.5 g/lit.
- ☆ Downy mildew: Spray combination of Metalaxyl + Mancozeb 2 g/l 3 sprays at 10 days interval.
- ☆ Black rot: Dip the seeds in 100 ppm Streptocycline for 30 minutes.
- ☆ Two sprays with 2 g/l Copper oxychloride + Streptomycin 100 ppm after planting and head formation.
- ☆ Yield Hills: 70 - 80 t/ha in 150 days, Plains - 25 – 35 t/ha in 120 days.

Chinese Cabbage, *Brassica compestris*

- ☆ Origin- China
- ☆ Family- Brassicaceae
- ☆ Originated from hybridization between turnip (*B.compestris spp.*) and pokchoa-(*B. compestris spp. Chinensis*)
- ☆ Pe- tsai– ***Brassica pekinesis***- 2n=2x=20
- ☆ Pak- choi- ***Brassica chinensis***- 2n=2x=20
- ☆ It is sensitive to water logging
- ☆ Seed rate- 500gmfor transplanting
- ☆ 2.5 kg/ha- Direct sowing
- ☆ Optimum temp. for growth- 15-20 ^{0}C
- ☆ Direct method for mostly sowing
- ☆ It can be harvested 40 days after harvesting
- ☆ Thinning 5-6 leaf stage very important operation in chinese cabbage
- ☆ It is cross pollinated
- ☆ Low O_2 (2 per cent), low CO_2 (2 per cent) - Improve self life of Chinese cabbage
- ☆ It is excellent for stire fry, pickling
- ☆ Pak choi-contain glycosinolates
- ☆ Var. Palampur Green, Chilli
- ☆ Pusa Sag. Wongbook- Suttons×Turnip
- ☆ Wong bok- Short and very chuky
- ☆ Solan selection
- ☆ Chinese season no- 1, non heading type (resistant to Alternaria leaf spot)
- ☆ Avg. yield 150-190 q/ha
- ☆ Spring January to April
- ☆ Tropical- April-July
- ☆ Autumn or main- August- September

Sprouting Broccoli/Broccoli: *Brassica oleracea* var. *italica* F: Cruciferae

- ☆ Chromosome number 2n=18
- ☆ Origin- Mediterranean region
- ☆ In USA, it is also known as Italian broccoli pointing to its Italian origin
- ☆ Inflorescence- Cymose
- ☆ It is more popular in India
- ☆ The word broccoli is derived from the Latin word *Brachium* meaning an arm or branch. Broccoli contains 3.3 per cent protein,
- ☆ High content of vitamin A and C and appreciable quantities of thiamine, riboflavin, niacin, calcium and iron.
- ☆ USA is largest producer of broccoli in world
- ☆ Optimum temp for proper growth 20-25 ^{0}C
- ☆ Temperature for heading 15-20 ^{0}C
- ☆ Head become loose when temp. will rise
- ☆ Different Types of Broccoli- Purple sprouting broccoli (Biennials, branched, purple spears, variable),
- ☆ Purple cape broccoli (Biennials, single purple heads, variable),
- ☆ White sprouting broccoli (Biennials, branched, white spears),
- ☆ Purple Sicilian broccoli (Single heading pale-purple heads also known in horticulture as purple cauliflower),
- ☆ Couve broccoli ("Roxo de Cabeca" Tall purple heading, sprouting type from Portugal), Calabrese (Green sprouting broccoli from Calabrian region.
- ☆ Now a world wide crop and virtually single heading),
- ☆ Black broccoli (Highly branched annual with dark green spears and sickle shaped leaves, from Rome area)
- ☆ Loamy soil is best
- ☆ 4-6 leaf old for transplanted
- ☆ Palam Samridhi high yielding variety 300-400gm weight
- ☆ Pusa KTS 1
- ☆ Male sterility: genetic male sterility
- ☆ Propagation mainly seed/vegetative propagation by cutting or tissue culture
- ☆ Seed rate 400-500 g/ha

- Sowing time- mid September- early November
- September- October in lower hills
- 4-6 weeks old seedling transplanting
- Spacing: 45x45 or 60x45 cm
- Broccoli is an Italian word from *latin brachium*
- Italy is the centre of its diversification
- It has about 130 times more vitamin A contain than cauliflower and 22 times more than cabbage
- It is a reach source of sulphorophane, anti-carcinogenic
- Properties reduce the risk of cancer
- Early cultivars are ready for harvest 40-50 days of transplanting
- Punjab broccoli green sprouting medium/green sprouting late are biennial
- Palam Samridhi, Pusa KTS-1- selection from exotic
- Palam Kanchan- Yellowish green heading broccoli
- Cultivars like Waltham 29, Green Mountain, Coastal Atlantic *etc.* are sensitive to buttoning and can be grown as an early or late crop.
- Extra Early: Southern Comet, Premium Crop and Laser
- Mid-season: Cosair, Excalibur and Emerald Corona
- Late: Late corona, stiff, kayak and Green Surf
- Pusa Broccoli Kt. Sel. 1: Developed through selection at IARI, Regional Station Katrain (Kullu Valley).
- Compact light green heads, weighing around 250-400g, plant height 40-50cm and ready for harvesting in 85-95days after transplanting.
- Palam Samridhi (DPGB I): Sprouting broccoli having compact green heads, free from yellow eyes and bracts, average head weight is 300-400g, plant branched.
- It also bears sprouts in the axil of leaves which adds to the total yield, ready for harvest in 85-90days after transplanting, average yield 150-200q/ha.
- Palam Haritika: Sprouting broccoli having dark green straight leaves, the main head formed deep inside the leaves.
- Late in maturity and gets 150 days for harvesting, free from yellow eyes, average yield 200-250q/ha
- Palam Kanchan: Yellow coloured heading type broccoli, leaves are large, green and straight with white midrib.

- ☆ Gets ready for harvesting in 140days, average yield 250-300q/ha.
- ☆ Palam Vichitra: Purple coloured heading type broccoli, medium foliage, open green leaves having green tinge, heads are firm and solid.
- ☆ Gets ready for harvesting in 115-120days, average yield 225-250q/ha
- ☆ Punjab Broccoli 1- This variety is developed by PAU, Ludhiana, matures in 65-70days, bears large number of spears, foliage dark green with smooth leaf surface, heads and spears are dark green with bluish tinge.
- ☆ Excessive use of cole crop resistant in swelling of thyroid gland and goitre disease
- ☆ Palam Samridhi- Recommended for sub tropical area
- ☆ Dry condition- Crop become fibrous
- ☆ Purple sicilian broccoli –Known as purple cauliflower
- ☆ Shallow root system
- ☆ Maturity-
- ☆ Early type- 40-50 days
- ☆ Mid type- 60-100 days
- ☆ Late type- 100 days of transplanting
- ☆ Bronzino- Purple variety,it heading broccoli type
- ☆ Yield- 5-15tonnes/ha
- ☆ Storage 3 ^{0}C, 88 per cent RH
- ☆ Degradation of ascorbic acid and chlorophyll was much slower under cool storage
- ☆ Highly cross pollination- 95 per cent
- ☆ Protogynous flower
- ☆ Wild Broccoli- *Brassica oleracea* var. *villosa*
- ☆ Broccoli refers to green buds and thick fleshy floral stalks arising from the main stem.
- ☆ A larger part of the produce in the USA goes to the freezing industry.
- ☆ It is also becoming popular in other parts of the country very rapidly.
- ☆ Sprouting broccoli is sometimes briefly called as broccoli though this name is also used for broccoli rape (the sprouts which develop on turnips) and for winter cauliflower (winter broccoli or heading broccoli).
- ☆ Morphologically, sprouting broccoli resembles cauliflower.
- ☆ The plant forms a kind of head, consisting of green buds and thick fleshy flower stalks.

- ☆ The sprouts in the axils of leaves develop strongly, especially after the removal of terminal bud or head.
- ☆ Both terminal head and the sprouts with bud clusters are consumed as human food.
- ☆ Hot weather results in development of small and thin leaves in the head. Generally, the plants of broccoli are hardy, can withstand fairly heavy frosts.
- ☆ It is generally biennial, while the annual cultivars are slightly sensitive to frost.
- ☆ Broccoli is slightly tolerant to acidic soils (pH 5.5-6.5).
- ☆ Purple and green variants occur though in USA, only the green types are grown.
- ☆ The purple types are fairly hardy and can be harvested during winter.
- ☆ The green sprouting broccoli is classified in accordance to their maturity i.e., early, medium and late cultivar.s.
- ☆ The demand is more for a firm head with small dark green buds while demand for the side shoots have decreased in recent years possibly due to difficulty in harvesting.
- ☆ The medium strains like Green Sprouting Medium take about 100days to maturity,
- ☆ While the late strains like Green Sprouting Late are biennials and may be harvested on the onset of winter or after its end.
- ☆ At present a large number of F1 hybrid cultivars are being marketed by the different seed companies of Japan, USA and Europe.
- ☆ The seed rate for raising seedlings to plant in one hectare area is about 300-400g.
- ☆ In plains, sowing is done in October whereas in mid hills it is done in September.
- ☆ In high hills, seeds are sown in month of May-June.
- ☆ The time of transplanting depends upon the climatic condition and the kind of variety.
- ☆ The sowing of an autumn crop is limited and sowing is undertaken in autumn to harvest them in late spring or early summer by over wintering them.
- ☆ Yield of broccoli ranges from 50-150q/ha depending on variety, time of planting and length of harvesting period.
- ☆ Seed yield of sprouting broccoli is about 5-6 q/ha.

Knol-Khol/Khol-Rabi *Brassica oleracea* var. *gongylodes*

- ☆ Chromosome number - 2n=18
- ☆ Family-Brassiceae/Cruciferae
- ☆ Origin- Mediterranean region
- ☆ Inflorescence Racemose
- ☆ Edible part- Knob
- ☆ Seedling ready in 30-40 day
- ☆ Spacing: 25×25cm, 25×30cm
- ☆ Average temperature for growth 15-20 °C
- ☆ Maximum minimum average 24 ^{0}C and 4.5 ^{0}C
- ☆ Sandy loamy soil is best, pH 5.5-6.8
- ☆ Best time for sowing October
- ☆ **Purple var. are more susceptible to pre mature bolting.** It is more popular in Kashmir- Sutton, earliest purple (Sutton)
- ☆ Early vienna
- ☆ King of the market north
- ☆ Early purple Vienna- Verma, pocha
- ☆ Pusa Virat, Palam tender knob.
- ☆ King of north- Puple spike is earliest variety
- ☆ Large green India
- ☆ Purple Vienna- Late variety
- ☆ White Vienna- Early variety- Most popular variety in India
- ☆ Seed rate - 1-1.5 kg/ha
- ☆ Average weight of knob- 200-250gm
- ☆ Yield- 12-30 tonnes/ha
- ☆ Seed yield- 700kg/ha
- ☆ Cross pollination 91 per cent
- ☆ Best time of sowing October

Kale (Collard); *Brassica olerecia* var. *acephala*. F- Brassicaceae

- Chromosome no.- 2n=2x=18
- Origin - Southern and western Europe
- Edible part - Fleshy stem
- Inflorescence- Racemose
- Very popular in Kashmir in every kitchen garden
- Highest Vitamin- A 9990 IU; Vitamin- C 187 mg/100g
- Tolerant to salt
- Introduced in 19th century
- It is a hardiest crop
- It can withstand -10 to -15 ^{0}C
- It is biennial but cultivated as annuals, it does not form head like cabbage
- Sandy loam soil is best for this, pH- 5.5 to 6.8
- Cross pollinated crops-83 per cent
- Lutein rich veg. crops 9.8- 13.4mg/100gm
- Dwarf variety having less than 40 cm height
- Dwarf green curled scotch
- Dwarf moss curled
- Mid tall (40-80cm height)
- Mass curled hamburger market
- Fiambarger market
- Karam sag is a mid-tall variety cultivated in Jammu and Kashmir
- Scotish and Siberian more suited for temperate region
- Commercially propagated by seeds
- Strongest saprophytic self –incompatibility
- Seedling ready -5-6 week
- Harvesting- 90- 120 DAP
- Seed rate 350- 400gm direct sowing 2-2.5kg/ha
- Spacing 45cm x 30cm, 60cm x 50cm
- Sowing time (August-October-November)
- Yield 100-250 q/ha.
- Storage life very short

Brussels Sprouts; *Brassica oleracea* var. *gemmifera*

- ✰ Chromosome number 2n=18
- ✰ Origin- Mediterranean region
- ✰ Inflorescence- Racemose
- ✰ Protogyny flower
- ✰ 72 per cent cross pollinated by honey bee or flies
- ✰ It is a minor crop
- ✰ Belgium brussel
- ✰ Edible part- Swollen axillary bud known as sprout or buttons and mini cabbage, round sprouts are about 2-5cm diameter
- ✰ **Tappin is necessary for increase harvesting time**
- ✰ Rich source of vitamin A/Ascorbic acid/Riboflavin/Calcium/Niacin
- ✰ **Morphological marker- Glossy foliage**
- ✰ Excessive application pot as import bitterness affects quality
- ✰ Slow growing and long bearing crops
- ✰ Spacing 70×45cm hybrid
- ✰ **Variety –**
- ✰ Hill ideal- Green colour
- ✰ Rubin- F_1 hybrid
- ✰ Dwarf- jade cross f_1 hybrid
- ✰ Tall- Amager market, Danish price
- ✰ Dwarf- Catskill
- ✰ Japanes- long is land
- ✰ Early morning
- ✰ **Rubin good flavour but low yield**
- ✰ It is a salt tolerant (C4-C8) but plant remain dwarf
- ✰ Grow under temperature range 7-24 ^{0}C highest yield 15-18 ^{0}C
- ✰ Temperature for seed germination around 12-16 ^{0}C
- ✰ When temperature is 16-20 ^{0}C and 8 ^{0}C higher temperature the growth is arrested
- ✰ Chilling temperature 5-8 ^{0}C for a period of 6-8 week from flower and seed
- ✰ It can be grown in light and loamy soil

- Sandy loam is Best for early type variety
- Clay loam is best for main season variety
- pH 6-6.5
- Dwarf 45-50 cm cultivars, tall upto 120cm
- 1st F_1 hybrid brussels sprout use in south India- avon cross
- Seedling are ready for transplanting after 4-5 week
- Planting time- September to October/June July transplanted
- October- November in north India
- Tapping method is more suitable for mechanical harvesting
- Harvesting- 90-120 days after sowing
- Seed rate -500 gm/ha
- Yield- 5-10 tonnes/ha
- 10 days internal from bottom of stalk
- Cultivation of broccoli/brussels sprout started in19 century
- Seed are sown 4-5 weeks before transplanting
- Genomic formula (AABBCCDEF) Broccoli/Brussels/Knol-Khol
- 1 gm seed contain 300-350 number of seed
- Self incompatibility is used in the development of hybrid variety
- Spacing for dwarf type 45×45cm, tall 45×60 cm
- Flavour and other dietery – Glucosinolates
- Mulching to contral of weed
- Cole crop are highly heterozygous and heterogeneous
- Fertilizer 25tonnes FYM, N-60kg, P_2O_5-200kg, K_2O- 40 kg

Chapter 3

Cucurbits

Cucumber:
Cucumis sativus L.; Cucurbitaceae 2n=14

- ☆ Progenitor of cucumber- ***Cucumis hardwickii***
- ☆ Day-neutral and C_3 plant
- ☆ They contain 0.4 per cent protein, 2.5 per cent CHO, 1.5mg iron, 2mg Vitamin C.
- ☆ ***KHULPI*** is a processed product prepared from cucumber.
- ☆ Temperate country extensively grown in glass house.
- ☆ Cucumber pickle acidity (0.6-0.8 per cent)
- ☆ Fruit is also used as (astringent and antipyretic).
- ☆ Fruits are used for constipation and jaundice.
- ☆ Triple cross hybrid of cucumber (TASKHA-2693)
- ☆ **1st sequenced vegetable crops.**
- ☆ Sex form monoecious.
- ☆ **Cucurbits has narrow genetic base (3-8 per cent polymorphism**).
- ☆ Bitterness in fruit due to cucurbitacin (**Metaxinic effect).**
- ☆ Chemically cucurbitacin is **(Tetracyclic triterpens**).
- ☆ **Aroma:** Non-adienal
- ☆ **Classification:**
- ☆ ***Cucumis metuliferus*** –(African horned cucumber)

- ***Cucumis anguria***- (West Indian Gherkin)
- ***Cucumis dispaceus*** (Hedge hogtesal-gourd).
- ***Cucumis hystrix***-Resistant todowny mildew, nematode, virus, gummy stem blight
- Green house: Cucumber has ability fruit set (Parthenocarpically)
- **1st F_1 hybrid Pusa Sanyog developed in India 1971.**
- Gynoecious female parent (Sub-tropical, Tropical, Temperate).
- Gene diversity values range (0.18-0.64).
- Introduce var. Japanese Long Green- Extra early long 45 day maturity, white skins introduce from japan
- Straight Eight- Early var. Introduce from USA
- Poinsette- Multi resistant var. develop from South Carolina.
- China Long Green.
- **HYBRIDS-**
- Himangi (Poinsett X Kalyanpur Ageti): Resistant to bronzing 100-110 days maturity.
- Phule Shubhangi (Poinsett X Kalyanpur Ageti) - High yield tolerant to powdery mildew.
- Solan hybrid F_1 hybrid: Most suitable for diara cultivation.
- **SELECTION**
- Sheetal
- K-75 high yield in temparete.
- Pant Sankar Kheera-1 (PCUC28 X PCUC8)
- Pusa Sanyog- Gynoecious line use
- Pusa Sanyog –(Japanese gynoecious X Green Long of Nepal)
- Private sector hybrid –Priya (Monoecious)
- Pant Kheera 01 (pcpu28)
- Pusa Barkha extra early var.
- Pusa Uday both season variety.
- Tokyo Long Green, Wins-Crimson Table Green, Poona Kheera
- Predominant hybrids: Gypsy (white fruit), Malini (green fruit), Harshini (green fruit)
- Narrow genetic base (3-8 per cent polymorphism)
- Genome size-880mbps

- ☆ Anthesis and temp.-5-7.30 am, temp. 13-18 °C
- ☆ Dehiscence 4: 30- 5 hrs.
- ☆ Isolation distance- 800m for foundation and 400m for certified.
- ☆ Seed yield: 400 kg/ha.; Fruit yield 700 kg/ha.; 500 seed/fruit,
- ☆ Test weight: Small seed -1000 seed (25g), large seed 1000 seed (33g).
- ☆ Spacing (1.5- 2.5 m) Row to row, hill to hill (60-90 cm)
- ☆ Poinsett (60-60 cm)
- ☆ **Gynoecious line treated by $AgNO_3$**
- ☆ Seed multiplication ratio =100
- ☆ **It is non-climacteric vegetable**
- ☆ Economic sex ratio 15: 1
- ☆ Seed rate (2.5-3 kg/ha.)
- ☆ Lime requirement (moderate)
- ☆ Respiration rate (moderate)
- ☆ Tolerance to soil acidity (moderate)
- ☆ Tropical gyonoecious line (304-C), (319-12), (338)
- ☆ Temperate gynoecious line (SR-551 F.W-12757)
- ☆ Soil: Sandy loam rich in organic matter with good drainage and pH range of 6.5-7.5.
- ☆ Soil temp. 15 °C for proper germination.
- ☆ **Pedigree method is most common method in cucumber.**
- ☆ **PGR:** Ethrel (150-200ppm) increase the no. of female flower
- ☆ GA_3 (10-25ppm) Increase flower, (MH 25-100ppm) fruit set.
- ☆ GA_3 (1500-2000 ppm) $AgNO_3$ (200-300ppm) induce the male flower on gynoecious cucumber.
- ☆ Chemical given (2-true leaf stage) and repeat (4-true leaf stage).
- ☆ Ethylene/auxin/acetylene/ethephone – Female flower increase.
- ☆ AVG-amino-venyl-thoxy-vinylglycine – Male flower increase.
- ☆ Storage 10 °C, RH- 85 per cent for (10-14 days).
- ☆ Male sterility- Cytoplasmic male sterility.
- ☆ Chilling injury as a physiological storage disorder of cucumber when exposed to below 15 °C.
- ☆ Open pollinated cultivar. of cucumber monoecious.
- ☆ Original form- Harmaphordite, pro-dominance, Andromonoecious.

- Optimum temp. 26 °C
- If temperature goes above 30 °C female flower production reduce, and male flower induce.
- Seed germination when day temp. 25 °C well drained loamy soil best.
- In Havier soil vines growth will be greater and fruit mature late.
- **Hybrid cucumber female: male (4: 2).**
- Cropping system - (relay cropping system and mix cropping).
- Preparation of field: Plough the field four times. Form long channels at 1.5 m apart.
- Season and sowing: Sow the seeds during June or January to April at 2.5 kg/ha after treating with *Trichoderma viride* 4 g or *Pseudomonas fluorescens* 10 g or carbendazim 2g/kg on one side of channel giving a spacing of 0.6 m between hills.
- Thin the seedlings to two per hill.
- Application of fertilizers - Apply FYM 40 t/ha as basal and 35 kg of N/ha at 30 days after sowing.
- Apply *Azospirillum* and Phospho-bacteria 2 kg/ha and *Pseudomonoas* 2.5 kg/ha along with FYM 50 kg and neem cake @ 100 kg before last ploughing.
- Form raised beds of 120 cm width at an interval of 30 cm and the laterals are placed at the centre of each bed.
- Sowing: Sow the seeds at an interval of 60 cm distance at the centre of the bed along the laterals.
- Spray pre-emergence weedicide like fluchloralin 1 kg a.i. or metolachlor 0.75 kg a.i./ha on third day of sowing.
- Fertigation: Apply a dose of 150: 75: 75 kg NPK/ha throughout the cropping period through split application for F1 hybrid in respect of phosphorous, 75 per cent has to be applied as a basal dose.
- Pests; Fruit fly: Fish meal and dichlorvos impregnated cotton are to be renewed once in 20 and 7 days respectively. Neem oil @ 3.0 per cent as foliar spray as need based. Do not use copper and sulphur dust, as these are phytotoxic.

Gherkin: *Cucumis sativus* var. *anguria*; Cucurbitaceae

- ☆ Soil: Well-drained sandy loam with a pH range of 6.0 to 6.8 is optimum.
- ☆ Seed rate: 800 gm per hectare.
- ☆ Sowing: Sow the seeds at 30 cm spacing on sides of the ridges with 2 seeds per hill after treating with *Trichoderma viride*@ 4 g or *Pseudomonas* @ 10 g or carbendaizim @ 2 g/kg of seeds.
- ☆ Preparatory cultivation: Apply 25 t/ha of FYM. Prepare ridges and furrows one metre apart.
- ☆ Manuring: Apply N - 150 kg, P - 75 kg and K - 100 kg/ha in 3 equal spittle, basal, three and five weeks after sowing.
- ☆ After cultivation: Earth up the plants 25 days after sowing. Provide support to plants as and when vines start trailing.
- ☆ Field preparation: Raise beds of 120 cm width at an interval of 30 cm and place the laterals at the centre of each bed.
- ☆ Minor pest: To control leaf miner, white fly, aphids and thrips spray Dimethoate 1.5 ml/l or Monocrotophos 1.5 ml/l or Malathion 1.5 ml/l.
- ☆ Diseases: Spray Carbendazim 0.05 per cent (0.5 g/l) to control diseases.
- ☆ Harvest: The crop is ready for harvest in 30-35 days.
- ☆ tender immature fruits are meant for canning the price of the produce is decided by the stage of maturity.
- ☆ Smallest fruit (stage 1) which will weigh approximately 4.0g (250 fruits per kg) will fetch the maximum price followed by stage 2 and stage 3.
- ☆ To maintain the grade the harvesting of fruits should be done every day.
- ☆ A day's break would end up with outsized or overgrown gherkin means loss to farmer.
- ☆ Avoid sharp sun and high temperature while harvesting.
- ☆ For this picking of fruits must be none in the very early morning or late evening.
- ☆ Harvest the fruits by retaining the stalk on the plant. Harvested fruits must be collected under shade.
- ☆ Flower head has to be removed from fruit. Water should not be sprinkled on harvested fruits at any stage.
- ☆ Even if there is surface water during harvest it should be dried by aeration.
- ☆ For collection of fruits jute bags alone have to be used and plastic bags should be totally avoided.

- The harvested produce should be transported to the factory on the same day before dusk.
- Leaving the gherkin unprocessed overnight would result in poor quality produce.
- Yield: 10 - 12 tonnes/ha in 90 days.

Watermelon: *Citrullus lanatus* (Thumb) Matsum and Nakai; Cucurbitaceae

- Chromosome number- 2n=22
- Origin- Tropical Africa (Kalahari dist.)
- Synonyms ***C. vulgaris***
- Edible part- Plancenta
- Largest producer China
- Tropical ratio for stiminate to pistillate- 7: 1
- Fruit contains 95 per cent water, 0.2 per cent protein, 0.3 per cent mineral, 3.3 carbohydrate/100gm
- Watermelon when grafted on bottle gourd incidence fusarium wilt
- Rich source of iron bitter substance- Cucurbitacin-E
- Seed colour white
- Cholesterol/sodium found in very low
- Personal size reduce watermelon developed in 21th
- Two types-
- Lanatus
- Citroides
- Small water melon done by J.M Crall in 1986. Minilee, kicklee
- Ancestor contains 160mg/100gm
- Anthracnose/downey mildew/powdery mildew resistant- Arka Manik
- TSS range 8-13 per cent
- Pigment- Anthocynin/lycopene
- Metallic sound- Harvesting shows immature
- Heavy sound- Show maturity
- 30 per cent yield loss due to weed
- Yellow tinge/brown spot it rests. on ground- indicate maturity
- Seed maturity and edible maturity coincide with each other
- **Variey- Introduction:**
- Asahi Yamato- Japanese introduction
- Fruit size inherited in polygene

- ☆ Sugar baby- 11-13 per cent TSS-USA introduction by Mr. Hardine 1956
- ☆ Pollination agent- Honeybee
- ☆ Improved Shipper- USA
- ☆ Storage temperature 12-13°C for 2-3 Weeks
- ☆ Test weight- 113gm
- ☆ 1st hybrid watermelon- Arka Jyoti
- ☆ Durgapur Lal- Leaves unlobed- Marker gene used
- ☆ Arka Manik- IIHR-21×Crimson Sweet
- ☆ Durgapur Meetha- Late Varieties
- ☆ Most popular cv. of watermelon Tri×313- Triploid
- ☆ Durgapur Keshar- late- yellow fleshed var.
- ☆ Special -1 red flesh and red seed- PAU Punjab
- ☆ Hybrid- Multiple disease resist. - Arka Manik
- ☆ In Rajasthan Mateera- grown in rainy season
- ☆ Arka Jyoti- IIHR20×Crimson Sweet- 11-12 per cent TSS
- ☆ Pusa Bedana- Triploid, (Tetra-2×Pusa Rasal)
- ☆ For transportation purpose- Charleston Gray
- ☆ Private sector- Madhur/Milan, All Jublient, All Producer, Red-N-Sweet
- ☆ Hallow heart and fibrous show over maturity
- ☆ Edible oil present in seed 45 per cent
- ☆ Seedling needed for transplanting one ha/area
- ☆ Yellow flesh cv. Yamato cream shin Yamato
- ☆ Climate-
- ☆ Exceess humidity will promotes disease-PM/DM/anthracnose
- ☆ The lack of inbreeding depression is due to small pollution size
- ☆ Good fruit quality- Dry weather
- ☆ Hybrid ration- 10: 1 female: male
- ☆ Temp.-35-40 °C during development
- ☆ Minimum 20-25 °C
- ☆ Hallow heart and fibrous flesh- sligh itly over maturity
- ☆ Excess irrigation- cracking
- ☆ Nov- January sown river bed
- ☆ North plain- Mid Feb in the garden

- ☆ Spacing 3.0-3.5 m
- ☆ P×P 60-90cm
- ☆ Cropping sytem- Relay system
- ☆ Maturity 90-120 days
- ☆ Seed rate 3-3.5kg/ha- small type
- ☆ 5kg/ha large type
- ☆ Germination would start with in 3 to 4 days
- ☆ Seed dressing with thirum controlled fungi effect
- ☆ Weed control- fluchlorelin 1.20 kg/ha
- ☆ PGR- increase the fruit and yield
- ☆ Boron 3-4ppm
- ☆ Average yield-30-40 tonnes/ha
- ☆ North Indian early sowing melon- river bed- Nov
- ☆ In pumpkin- ***Cucurbita lundelliana***- good source of resistant to powdery mildew
- ☆ Spray hair persisting on the skin of bottle gourd/round melon/tinda
- ☆ Biological control- *Entomophthora cormata*
- ☆ Useful parasite on cucurbits aphid
- ☆ Vit. A and C are sensitive to oxidative degeneration
- ☆ Thiamine can be destroyed by heat
- ☆ Sulfering and riboflavin by light
- ☆ Centre of origin- India
- ☆ Edible part – Mesocarp
- ☆ Largest producer China
- ☆ Fruits contain 95 per cent water, 0.2 per cent protein, 0.3 per cent mineral, 3.3 per cent CHO.
- ☆ Potassium contain- 160 mg/100gm.
- ☆ Shallow rooted.
- ☆ Ancestor of watermelon- ***Citrullus colocynthoides.***
- ☆ TSS rangc 8-13 per cent, pigment anthocyanin and lycopene.
- ☆ Metallic sound- Harvesting shows (immaturity)
- ☆ Heavy sound – Harvesting shows (maturity).
- ☆ 30 per cent yield loss due to weeds.

- ☆ Small water melon done by J.M crall (1986), minilee, kicklee.
- ☆ Cholesterol and sodium found in very low
- ☆ Two types of watermelon: 1- lanatus; 2- citroides.
- ☆ Rich source of iron.
- ☆ Watermelon when grafted bottle gourd cv. (***rensi***) they incidence fusarium wilt.
- ☆ ***C. Vulgaris, C. colocynthi, C. ecirrhosus***- Bitter substance (cucurbitacin E) tetracyclic triterpence.
- ☆ ***Citrulus naudinious*** both cucurbitacin E and B
- ☆ Multi resistance var. of watermelon- Arka Manik.
- ☆ Yellow tinge and brown spot it resistant on ground indicate maturity.
- ☆ Seed maturity and edible maturity coincide with each other.
- ☆ Fruit size – 6
- ☆ Mini watermelon (less 4kg), Icebox (4.0-5.5kg),Small or pee wee (5.5- 8.0 kg) medium (8.0-11 kg), large (11-14.5 kg), Giant (more than 14 kg),
- ☆ Asahi Yamato –Introduce by Japanese.
- ☆ Fruit size inherited in polygene Sugar Baby and Arka Manik.
- ☆ Hybrids: Arka Jyoti and Pusa Bedana.
- ☆ Soil: Sandy loam rich in organic matter with good drainage and pH range of 6.5-7.5.
- ☆ Season and sowing: Treat the seeds with *Trichoderma viride* @ 4 g or *Pseudomonas fluorescens* 10 g or carbendazim 2g/kg of seed.
- ☆ Thin the seedlings 2 per hill 15 days after sowing.
- ☆ Preparation of field: Plough the field to a fine tilth and form long channels 2.5 m apart.
- ☆ Irrigation: Irrigate the field before dibbling the seeds and there after once a week.
- ☆ Application of fertilizers: Apply FYM 20 t/ha, P 55 kg and K 55 kg as basal and N 55 kg/ha 30 days after sowing.
- ☆ Fertigation Schedule: Recommended Dose: 200: 100: 100 Kg/ha
- ☆ Apply *Azospirillum* and Phosphobacteria @ 2 kg/ha and *Pseudomonoas*@ 2.5 kg/ha along with FYM 50 kg and neem cake 100 kg before last ploughing.
- ☆ Nursery for watermelon can be prepared either with polythene bags of 200gauge, 10 cm diameter and 15 cm height size or through protrays under protected nursery.

- In polybag nursery, fill the bags with 1: 1: 1 ratio of red soil, sand and farmyard manure mixture.
- Use pro-trays, each having 98 cells for raising seedlings.
- Transplant about 12 day's old seedlings in the main field.
- Drip irrigation: Install drip system with main and sub-main pipes and place the in line lateral tubes at an interval of 1.5 m.
- Place the drippers in lateral tubes at an interval of 60 cm and 50 cm spacing with 4 LPH and 3.5 LPH capacities respectively.
- Field preparation: Raise beds of 1.2 m width and 30cm height for sowing.
- Spray pre-emergence weedicide (Pendimethalin @ 1kg a.i/ha) just before planting. Plant the seedlings in the holes made at 60 cm distance.
- Spray ethrel 250 ppm (2.5 ml/10 litre of water) 4 times at weekly intervals commencing from 15 days after sowing. Hoeing and weeding thrice.
- Fruit fly: The fly population is low in hot day conditions and it is peak in rainy season.
- Expose the pupae by ploughing. Use 20 x 15 cm polythene bags, fish meal trap with 5 g of wet fish meal and 1 g of dichlorvos in cotton, 50 traps are required per hectare.

Muskmelon: *Cucumis melo* L; Cucurbitaceae

- Origin- Tropical Africa
- Chromosome number 2n=2X=24
- Common name- Whole some food/cantaloup/winter melon
- Size of chromosome- 1.06-1.88µm
- TSS range- 11-17 per cent
- ***Cucumis melo var.- common makino***-Oriental picking melo
- ***C. melo var. anguina***- Pickles- West India
- Chines white cucumber genome size- 450-500 mbps
- ***C.melo. Momordica*** (Snap melon)/phoot/phoonte
- ***C.meloindorus***- Musky flavour
- ***C.meloconomon***- High sugar contient
- ***Cucumis anguria***- West Indian gherkin
- ***C.melo- contalupensis***- Aromatic flavour
- Jeffery 1980 classified- 2 sub genera
- ***Cucumis sativus, Cucumis hystrix***
- **Anguria group**- Dioceous/monoecious/andromonoecious/***C. anguria/C. dipsaceus/C. prophetarum***
- **Matulifrus group**- Monoecious/annual with red spiny fruit/***Cucumis metuliferus*** melo group- Monoecious/andromonoecious/perennial
- ***C. anguria/C.ficifolius*** are tolerance to aphid/PM
- Anthesis takeplace- 5: 30-6: 30 pm at tem- 22-29 ^{0}C
- Dehiscence take place before anthesis
- Fruit set higher when muskmelon used as the pollen parent
- Higher fruit set is obtained under natural pollination as compared to hand pollination
- Andromonoecious nature of muskmelon involve the tedium of muskmelon of perfect flower in the female parent for hybrid seed production
- Heterosis in muskmelon by using andromonoecious var.
- Pusa Rasraj using muskmelon monoecious sex as female parent
- Polyploidy breeding in muskmelon early as in 1953 by using colchicine
- Fruit set in monoecious line of ***C. melo*** 29-40 per cent

- ☆ PGR- male flower induce-Silver thiosulphate
- ☆ Increase fruit yield- ethereal @ 250ppm
- ☆ Pre cooling of melon is important post harvest practice
- ☆ Hydrocooling is the most efficient and rapid cooling method for cantaloups
- ☆ C.callasus are highly tolerant to fruit
- ☆ Snap melon- resistant to DM/PM/CGMMV
- ☆ Musk melon- Pusa Sandar var.
- ☆ Snap melon and musk melons are intercrossable
- ☆ Male sterility has been commercially exploited for the hybrid seed production – muskmelon
- ☆ Sarda- Afghanistan grown var. is available in India market in oct- nov
- ☆ Classification of muskmelon var. given by- Nandin
- ☆ ***C. melo var. agrestis***
- ☆ ***C. melo var. ruticulatus***- Netted melon
- ☆ ***C. melo var. flexuosus***- Snake or long melon
- ☆ ***C. melo chito***- Mango/lemon melon
- ☆ ***C. melo var. dudaim***- Ornamental purpose
- ☆ ***C. melo var. momordica***- Snap melon muskmelon is harvested at full slip stage- netting except Hara madhu
- ☆ **Frost susceptible**
- ☆ Excess humidity- Promotes- PM/DM/Anthracnose
- ☆ Good fruit quality/sweetness/in muskmelon- during dry weather
- ☆ 35-40 ^{0}C during fruit development
- ☆ Cool night and warm day are ideal for accumulation of sugar in fruit
- ☆ Maturity is hasten, if night are warm
- ☆ Average temp 30-35 ^{0}C for growth
- ☆ Max-40 ^{0}C
- ☆ Minimum- 20-25 ^{0}C
- ☆ Most of cucurbits germinate when temp. will be above 25 ^{0}C
- ☆ Well drained loamy soil is best
- ☆ Soil pH- 6-7.0
- ☆ More tolerant to soil acidity
- ☆ Soil moisture 10-15 per cent

- ☆ Cucumber- maximum and minimum temp. 25 and 18-25 ^{0}C respectively
- ☆ Seed rate 1kg/ha deebling method
- ☆ OP- 2.5-3.75 kg/ha
- ☆ 1.5-2.0 kg/ha general
- ☆ Spacing: R×R 1.5-2.5 cm
- ☆ Hill- 60-120 cm
- ☆ Earlier soil temp should not be go below- 10 ^{0}C, otherwise there would not be any seed germinate
- ☆ Isolation distance 800 M, 400 M
- ☆ Seed treatment- ethephone 480mg/lit. for 24 hours
- ☆ Seedling are ready for transplanted- feb end- 1st week of march
- ☆ Pitcher irrigation is best
- ☆ Mixed cropping of melon
- ☆ Multi aroma found in ripend muskmelon
- ☆ Max no. of fruit with highest TSS is produce between 9-12 months node of the main stem
- ☆ Weed control- fluchloralin- 1.2kg/ha pre plant
- ☆ Fruit maturity indicate full slip stage
- ☆ Climatric fruit
- ☆ Solmon orange colour rich- carotenoid pigment
- ☆ Storage- 0°C at 80-90 per cent RH 2-3 week
- ☆ Test weight- 25gm
- ☆ **Var.- selection**
- ☆ Durgapur Madhu- extra early var.
- ☆ 1st hybrid- Punjab hybrid-1
- ☆ Arka Ranjhas- mid var.
- ☆ Arka Jeet- early var./excellent floured high Vit C
- ☆ Hara Madhu- TSS 12-15 per cent
- ☆ Seed yield- 300-500kg/ha
- ☆ Contaloupe 45 (PMR-45)- powdery mildew resistance var.
- ☆ Hybrid-
- ☆ Pusa Sharbati- Kutana×Contalonpes- early var.
- ☆ Suitable for river bed cultivation

- Pusa Rasraj- Monoecious-3×Durgapur Madhu
- Hissar Madhu- Pusa Sharbati×75
- Punjab Rasila- Resistance to PM/DM
- Kashi Madhu-
- M.H.Y.S- Durgapur Madhu× Hara Madhu
- Public sector hybrid-
- F1 hybrid- Pusa Rasraj- Developed using monoecious sex form as female parent- 11-12 per cent TSS yield 17.3 tonnes/ha
- Arka Sheetal is an improved cultivar of long melon
- Lucknow Safeda, Deepti is most common hybrid
- Drainage loamy soil preferred
- Snap melon take about 27-28 days from anthesis to harvesting
- Pusa Sandar- improved var. of snap melon
- Winter melon var.- Golden Beauty, Santaclaus, Honey Dew
- Muskmelon var. susceptible to frost
- Soil moisture should be above 10-15 ^{0}C permanent wilting point
- Seed rate 3-7 kg/ha
- 2.5-3.5 kg/ha long melon
- Germination in muskmelon 3-4 days
- Higher dose of N_2- higher temp. - Promotes maleness
- Critical stage of irrigation-Flowerbud development
- Pither irrigation save water- 40 per cent
- 1st plant growth regulator give at 2 true leaf stage
- 2nd plant growth regulator give at 4 true leaf stage
- 2-3 days store under ambient condition
- Plant size 1-10 m
- Fruit weight 10gm-10kg
- Fruit flesh TSS 3-18 per cent
- Fruit flesh acidity- pH 3-7
- Powdery mildew resist governed by single dominant gene
- 1st gynocious line in muskmelon developed by Peterson 1983
- 1st hybrid national level- Punjab hybrid 1984
- 1st recessive male sterility line- reported by Bohn and Whitekar 1949

- Min. standard TSS for muskmelon-10 per cent
- Leading producer of Garkin in India KN>TN
- Economic sex ratio 15: 1
- Primitive sex form hermaphrodite
- Adavance sex form monoecious
- Soil: Sandy loam rich in organic matter with good drainage and pH range of 6.5-7.5.
- Season of sowing: November to February.
- Sow the seeds @ 3.0 kg/ha after treating with *Trichoderma viride* @ 4g or *Pseudomonas fluroscens* @ 10 g or carbendazim@ 2g/kg of seed on one side of the channel giving a spacing of 0.6 m between hills.
- Thin the seedlings after 15 days, to maintain two seedling per hill.
- Preparation of field: Plough the field to a fine tilth and form long channels at 2.5 m apart.
- Irrigation: Irrigate the field before dibbling the seeds and thereafter once in a week.
- Application of fertilizers - Apply FYM 20 t/ha, NPK 40: 60: 30 kg/ha as basal and N @ 40 kg/ha 30 days after sowing.
- Nursery preparation: Nursery for muskmelon can be prepared either with polythene bags of 200gauge, 10 diameter and 15 cm height size or through protrays under protected nursery.
- Use protrays, each having 98 cells for raising seedlings.
- Transplant about 12 days old seedlings in the main field.
- Drip irrigation: Install drip system with main and sub-main pipes and place the inline lateral tubes at an interval of 1.5m.
- Place the drippers in lateral tubes are at an interval of 60 cm and 50 cm spacing with 4 LPH and 3.5 LPH capacities respectively.
- Sowing: Around 250 protrays are required for the production of 23,334 (22,223 + 5 per cent) seedlings, which are required for one hectare at spacing of 1.5 m x 30 cm in a raised bed single row system.
- Raise beds of 120 cm width at an interval of 30 cm and place the laterals at the centre of each bed.
- Direct sowing or transplanting is done at a spacing of 1.5 m along the laterals and 30 cm interval in the raised bed single row system, using ropes marked at 30 cm spacing.
- Fertigation: Apply a dose of 200: 100: 100 kg NPK/ha throughout the cropping period through split application.

- Fertigation schedule-Muskmelon Recommended dose: 200: 100: 100 Kg/ ha
- 75 per cent RD of Phosphorus applied as superphosphate = 75 x 6.25 = 469 kg/ha.
- The fly population is low in hot day conditions and it is peak in rainy season.
- Use 20 x 15 cm polythene bags, fish meal trap with 5 gm of wet fish meal and 1 g of dichlorvos in cotton, 50 traps are required per hectare.
- Fish meal and dichlorvos impregnated cotton are to be renewed once in 20 and 7 days respectively.
- Neem oil @ 3.0 per cent as foliar spray as need based. Do not use copper and sulphur dust, as these are phytotoxic.
- White fly: Spray neem seed kernel extract 5 per cent. Yield: 20 t/ha in 120 days.

Pumpkin: *Cucurbita moschata* Poir.; Cucurbitaceae

- Chromosome number 2n=40
- Origin- tropical America
- Monoecios plant
- Pumpkin festival celebrated in USA
- Genomic structure of cucrbita moschata is Aaabb, amphidiploid
- Vit. A- 1600IU
- Yerusseri product – popular in kerala
- Largest fruit weight- 857kg/ha
- Popular summer squash- zucchini
- *Cucrbita maxima* is grown mostly on the hills
- Proginator of summer squash-***Cucrbita fratena***
- Fig leaves gourd or Malabar gourd ***C. ficifolia***
- Buffalo gourd ***C. foetidissima***
- Germination starts- 7-8 days
- Petted gourd- ***C. lundelliana***- resist to (P.M)
- No inbreeding depression in pumpkin
- Flower of pumpkin leaves are more nutrativevalaue than fruit
- Plant grown successfully temp range- 25-24 ^{0}C
- Channel and hill system of pumpkin cultivation is most useful for higher yield
- Chief pollinator- honey bees
- Male and female ration 1: 5
- Maturity 75-180 day
- Fruit drop is common problem in pumpkin
- Arka Chandan- highest b-carotene -3331 IU/100gm
- Pusa Hybrid- sel. 107×sel. 124
- Pusa Vishwas- Local selection line-CM-10
- Pusa Vikas- rich in vit A- small fruit- dwarf to dwarf type
- CM14, CO1, CO2, Kashi Harit
- Ambali, Arka Suryamukhi- resistant to fruit fly
- Solan Badami- good keeping quality
- Fruit maturity 160-220 DAS

- Crop duration 120-140 day
- Kashi Harit (NDPK24×PKM) – Pedigree Method
- Fruit can kepts10-12 ^{0}C for 6months
- Sandy loam rich in organic matter and with good drainage.
- The pH ranges from 6.5-7.5 is found ideal.
- Season and sowing: June- July and December- January.
- Soak the seeds in double the quantity of water for 30 minutes and incubate for 6 days.
- Sow the seeds (3 seeds/pit) treated with Azospirillum just before sowing and thin the seedlings to two per pit after 15 days.
- Seed rate: 4.5 kg/ha.
- Spacing: 2 m x 2 m. Pit size 30 cm x 30 cm x 30 cm.
- Application of fertilizers: Apply 10 kg of FYM (20 t/ha) and 100 g of NPK 6: 12: 12 mixture as basal and 10 g of N per pit after 30 days of planting.
- Apply *Azospirillum* and Phosphobacteria @ 2 kg/ha such and ***Pseudomonas*** 2.5 kg/ha along with FYM 50 kg and neem cake @ 100 kg before last ploughing
- Spray Ethrel 250 ppm (2.5 ml per 10 l of water) four times at weekly intervals commencing from 10 to 15 days after sowing.
- Nursery raising: use 12 days old healthy seedlings obtained from shade net houses for planting. Raise the seedlings in pro-trays having 98 cells.
- Use well decomposed coco-peat as medium. Sow one seed per cell.
- Water regularly twice a day.
- Fertigation: Apply a dose of 60: 30: 30 kg NPK/ha throughout the cropping period through splitapplication. Apply 75 per cent of the phosphorus as superphosphate as basal dose.
- Beetles and caterpillars: Spray Dichlorvos 76 per cent EC @ 6.5 ml/10 l or ***Trichlorofen*** 50 per cent EC @ 1.0 ml/lit.
- Fruit fly: Use polythene bags fish meal trap with 5g of wet fish meal + 1 ml di-chlorphos in absorbent cotton.
- 50 traps are required per ha, fish meal + dichlorvos soaked cotton are to be renewed once in 20 and 7 days, respectively).
- Downy mildew: Spray Mancozeb or Chlorothalonil 2 g/l twice at 10 days interval.
- Yield 20-30 tonnes/ha in 140 days.
- Grade Specification Clean and glossy, bright appearance, peel changes from green to yellow, yellowish flesh colour

Bitter Gourd: *Momordica charantia* L.; Cucurbitaceae

- ☆ Chromosome number 2n=22
- ☆ Family- Cucrbitaceae
- ☆ Origin- Tropical asia
- ☆ Proginator- ***M. charantha var. abbreviata***
- ☆ Common name- Balsom pear/Bitter cucumber/Karela
- ☆ Length of chromosome range 1-2 μm, 6 chromosome pair
- ☆ Day neutral plant
- ☆ Fruit are covered with blunt tubercles
- ☆ Fruits are rich in iron, Vit. A, B and C
- ☆ Isolation distance- 800m foundation, 400m certified
- ☆ Cucrbitacin- Bitter glucoside
- ☆ 1000seed- 175gm
- ☆ Non bitter fruit- Balsam apple
- ☆ Other species- ***Momordica dioca***- Spine gourd kakarol 2n=28
- ☆ ***M. cochinesis***- sweet gourd 2n=28
- ☆ Pollination can perform daily in morning hrs
- ☆ Plant is a monoecious annual climber
- ☆ Flower- Axillary solitary
- ☆ Fruit is pendules/fusiform
- ☆ Seed germination inhance by GA3 @25-50ppm
- ☆ Anthesis and anther dehiscence occurs early in the morning
- ☆ Inheritance the female; male ratio MH @ 50-100ppm/CCC @ 50-100ppm
- ☆ Bitter taste is due to- Momordicin
- ☆ Optimum temp-30-35 ^{0}C
- ☆ Minimum temp 18 ^{0}C then growth is low and poor quality
- ☆ When temp is above 40 ^{0}C male flower produce poor production of female flower
- ☆ Seed germination optimum temp- 25-35 ^{0}C
- ☆ Seed yield- 6 q/ha
- ☆ Yield- 100-150 q/ha
- ☆ Seed rate- 4.0 -5.5 kg seed/ha

- Transplanting in pit- 2.5-3.0kg/ha
- Short day treatment slighty reduce the production of staminate flower and increase the no of female (pistilate) flower
- Colour white/pink
- Long day treatment increases the production of stiminate flower and developed of pistilate flower
- Anthesis and dehiscence- early in the morning
- Gynoecious line- developed by- T.K Behera IARI
- Selfing and crossing- Perform morning
- Sandy loam/silt loam soil are best
- pH 6.5-7
- Jan.-feb. in north India
- April- July- hill area
- Seed treatment- Thiram @2g/kg of seed
- PGR- MH 50-150 ppm/CCC 50-100ppm- increase female/male ration
- High concentration of 200ppm CCC reduce male/female flower
- Ethrel-25ppm increase female flower
- High length of GA_3- low male and female flower
- GA_3 at 25mg/use- give highest no. seed/fruit
- Flowering start- 45-55days
- 1st pickling 60-70days planting
- Use of asapirillium and photo bacteria- good quality fruit
- Training system in South India- Bower system
- Gynoecious in bitter gourd is controlled by single recessive gene
- Arka Harit- Spindle shape
- Gynocious line- Gy263B given by Dr. D Ram
- Pusa Vishes- dwarf vine variety, stuffing and dehydration
- Suitable for spring and summer season
- Pusa Do Mausami- selection var.
- Preethi- Whitc flower
- Priyanka- White flower
- Priya- Fruit white, extra long fruit 40cm
- Kalyanpur Baramasi- Thin and topering type var.

- Arka Anupam
- Small fruit var.- 7.5-10cm, Pusa Do Mausami, Pride of Surat, Pride of Gujarat, Small Grace
- Konkan Tara- Suitable for- export, good keeping quality
- Kalyanpur Sona- Stuffing var.
- Harkari – Sel. from local- long fruit- 15-20cm
- Kashi Urvarshi- IC-85650B×IC44435A, both season
- Pant Kerela- Highly resistant to red pumpkin beetle
- Coimbatore long white- extra long 45cm, white colour
- RHR, BGH-1- F_1 hybreed
- MDU-1- Mutant var.- Mutation of MC103 material
- Fruit fly and Baramasi tolerant var.- Kalyan Baramasi
- Phule Green Gold- Green Long×Delhi local- Tolerant to DM
- Red pumkin beetle resist- Pant Karela-1
- Heterosis in bittergourd wart 1st studied by Palandsinghin 1946
- Gynoecious line in bittergourd was developed by T.K. Bahera, IARI New Delhi
- Seed protein ricine of bitter gourd inhibited growth of immune virus HIV in human cell culture
- Recessive gene desined by- Gy-1
- Sandy loam rich in organic matter with good drainage and pH range of 6.5-7.5.
- Season and sowing July and January.
- Sow the seeds (3 seeds/pit) treated with *Trichoderma viride* @ 4 g or *Pseudomonas fluorescens* @ 10 g or carbendazim @ 2 g/kg of seeds and thin the seedlings to two per pit after 15 days.
- Preparation of field: Plough the field to fine tilth. Dig pits of 30 cm x 30 cm x 30 cm size at 2 x 1.5 m spacing and form basins.
- Irrigation: Irrigate the basins before dibbling the seeds and thereafter once in a week.
- Application of fertilizers Apply 10 kg of FYM (20 t/ha) and 100 g of NPK 6: 12: 12 mixture per pit as basal and 10 g of N per pit 30 days after sowing.
- Apply ***Azospirillum a***nd Phosphobacteria 2 kg/ha and ***Pseudomonas*** @ 2.5 kg/ha along with FYM 50 kg and neem cake @ 100 kg before last ploughing.
- Fertigation Schedule- Recommended dose: 200: 100: 100 kg/ha

- ☆ Quality seedling production: Bitter gourd is a direct sown vegetable but polythene bag nursery is more advantageous to get early marketing and to avoid more gap filling.
- ☆ Use 200gauge poly bags of 10 cm diameter x 10 cm height for sowing the seeds.
- ☆ Transplant about 15 day's old seedlings to the main field.
- ☆ Drip irrigation: Install drip system with main and sub-main pipes and place the inline lateral tubes at an interval of 1.5m.
- ☆ Place the drippers in lateral tubes at an interval of 60 cm and 50 cm spacing with 4 LPH and 3.5 LPH capacities respectively.
- ☆ Field preparation: Form the raised beds of 120cm width (120 cm) and spread the lateral tubes in the centre of each bed.
- ☆ Irrigation is done in the beds by operating the drip system continuously for 8-12 hrs.
- ☆ Spray pre-emergence herbicide like Pendimethalin @ 1 Kg a.i./ha just before planting. Planting or sowing is done at the holes made at 2 m distance.
- ☆ Fertigation: Apply a dose of 200: 100: 100 kg NPK/ha throughout the cropping period through split application.
- ☆ After cultivation Hoeing and weeding thrice. Provide stakes for the plants to reach the pandal (2 m).Spray Ethrel 100 ppm (1 ml dissolved in 10 litre of water) four times from 15th day after sowing at weekly intervals.
- ☆ Pests; Mites: Spray dicofol 18.5 per cent SC @ 2.5 ml per lit.re of water Aphid: Spray Imidachloprid @ 0.5 ml/lit. along with sufficient quantity of stickers like Teepol, triton X100, apsa *etc.*, for better adhesion and coverage.
- ☆ Powdery mildew: Spray Dinocap 1ml/l or Carbendazim 0.5 g/l.
- ☆ Downy mildew: Spray Mancozeb or Chlorothalonil @ 2 g/l twice at 10 days interval.
- ☆ Grade Specification 20-25 cm long green fruits with short neck and tubercles are preferred

Bottle gourd: *Lagenaria siceraria* (Mol) Standl; F.- Cucurbitaceae

- ☆ Chromosome number- 2n=22
- ☆ Origin- South Africa
- ☆ Common name- Calabash/White flower/Zuccz melon
- ☆ 0.2 per cent protein, 2.9 per cent CHO, 0.5 per cent fat, 11mg Vit. C
- ☆ Shallow rooted crop
- ☆ Recessive gene is responsible for long type fruit
- ☆ Leaf is a very good medicine for curring jaundice
- ☆ Pulp is good for- cough/night blindness
- ☆ It is suitable for cultivation in dry area
- ☆ Kofta is most popular preparation dish
- ☆ Fruit colour for exported light green to dark green
- ☆ Pusa Summer Prolific Long/Pusa Hybrid-3/Samrat
- ☆ Good export var.
- ☆ It is cross pollinated crop
- ☆ Pure line selection is most important breeding method
- ☆ Flower colour- White
- ☆ Optimum temp for growth and development 24-27°C
- ☆ Soil temp for good temp. 18-22°C
- ☆ Soil moisture is important for rapid growth- 10-15 per cent
- ☆ It is highly sensitive to photoperiod
- ☆ High rainfall- prolonged cloudness result in a highr incidence of disease reduce the yield
- ☆ Season- both- Summer/rainy season
- ☆ Summer dec.-jan.Rainy June-july
- ☆ Propagation- Seed
- ☆ Seed large white in colour- 450-500seed/100gm
- ☆ Seed rate- 3-6kg/ha
- ☆ Seed treatment- 60ppm- succinic acid for 12hrs
- ☆ Bottle gourd is trained on-Bower sytem give high yield
- ☆ Yield- 80tonnes/ha

- ☆ Export quality fruit 30-40cm long- liquid green colour
- ☆ Male: female flower ratio 5: 5
- ☆ GA_3 lower concentrated increase- female flower
- ☆ Drip irrigation 48 per cent yield more then furrow irrigation
- ☆ GA_3 higher- concentration increase male female
- ☆ Weed management- fluchloraline 2.kg/ha
- ☆ Without Croock neck- Arka Bahar
- ☆ PGR-
- ☆ Fruit set increase- Ethrel 100-150ppm
- ☆ M.H 400 ppm- Promotes female flower and increase yield
- ☆ Pusa Hybrid-3 Early picking var.- Pusa Naveen× Selection-48
- ☆ Harvesting- 60 days after sowing
- ☆ Cowleg is multiviral resistant variety
- ☆ Average yield 20tonnes/ha
- ☆ F_1 hybrid 40-50 tonnes/ha
- ☆ 1st public sectore hybrid- Pusa Meghadoot- PSPB×Selection
- ☆ Cross pollinated cross- Pusa Hybrid-2 Club shaped fruit
- ☆ Female male ratio-2: 1 or3: 1
- ☆ Kashi Ganga- IC92465×DVBG151
- ☆ Pinching of male female on female plant is commercially practiced for hybrid seed
- ☆ Punjab Komal- tolerant to cucumber- mosaic virus
- ☆ Pusa Majari- Round type- F_1 hybrid (PSPR×Selection-11)
- ☆ Pusa Summer- prolific round
- ☆ Pusa Sandesh- Both season
- ☆ Pusa Samridhi- Non crook neck var.
- ☆ Pusa Santusti- Pear shaped fruits.
- ☆ Pusa Naveen- Non croock neck var.
- ☆ Arka Bahar- Non croock neck var., export var.
- ☆ Samrat- Height yield
- ☆ Sandy loams rich in organic matter with good drainage and a pH range from 6.5 to 7.5.
- ☆ Season and sowing: July and January. Sow the seeds (3 seeds/pit) treated with ***Trichoderma viride*** @ 4 g or *Pseudomonas fluorescens* 10 g or

carbendazim 2 g/kg of seeds and thin the seedlings to two per pit after 15 days.

- Preparation of field: Plough the field to fine tilth.
- Dig pits of 30 cm x 30 cm x 30 cm size at 2.5 x 2 m spacing.
- Irrigation: Irrigate the field before dibbling the seeds and thereafter once a week.
- Application of fertilizers: Apply 10 kg of FYM (20 t/ha) and 100 g of NPK 6: 12: 12 mixture as basal and 10 g of N per pit 30 days after sowing.
- Apply *Azospirillum* and Phosphobacteria 2 kg/ha and *Pseudomonas* 2.5 kg/ha along with FYM 50 kg and neem cake @ 100 kg before last ploughing.
- Fertigation Schedule-Recommended Dose: 200: 100: 100 Kg/ha
- Drip irrigation: Install drip system with main and sub-main pipes and place the inline lateral tubes at an interval of 1.5m.
- Place the drippers at an interval of 60 cm and 50 cm spacing with 4 LPH and 3.5 LPH capacities respectively.
- Form raised beds of 120cm width and place laterals in the centre of bed.
- plant 15 days old healthy seedlings raised in shade net houses.
- Raise the seedlings in protrays having 98 cells or in polythene bags.
- Transplant about 15 day's old seedlings in the main field.
- Apply a dose of 200: 100: 100 kg NPK/ha throughout the cropping period through splitapplication.
- Mites: Spray dicofol 18.5 per cent SC @ 2.5 ml per lit.re of water
- Aphid: Spray Imidachloprid @ 0.5 ml/lit. along with sufficient quantity of stickers like teepol, triton X100, apsa *etc.*, for better adhesion and coverage.
- Powdery mildew: Spray Dinocap 1 ml/l. or Carbendazim 0.5 g/l or Tridemorph l ml/l.
- Downy mildew: Spray Mancozeb or Chlorothalonil 2 g/l. twice at 10 days interval.
- Yield: 15 – 20 t/ha in 135 days.

Ridged Gourd: *Luffa acutangula* Roxb. Cucurbitaceae

- Soil: Sandy loam rich in organic matter with good drainage
- pH ranges from 6.5-7.5.
- Season of sowing: July and January.
- Seed rate: 1.5 kg/ha.
- Sow the seeds (3 seeds/pit) treated with *Trichoderma viride* @ 4 g or *Pseudomonas fluorescens* @ 10g/kg of seeds and thin the seedlings to two per pit after 15 days.
- Preparation of field: Plough the field to fine tilth. Dig pits of 30 cm x 30 cm x 30 cm size at 2.5 x 2 m spacing and form basins.
- Irrigation: Irrigate the basins before dibbling the seeds and thereafter, once a week.
- Application of fertilizers Apply 10 kg of FYM, 100 g of NPK 6: 12: 12 mixture as basal per pit and N @ 10 g per pit 30 days after sowing.
- Apply *Azospirillum* and Phosphobacteria @2 kg/ha and *Pseudomonas* 2.5 kg/ha along with FYM 50 kg and neem cake @ 100 kg before last ploughing.
- Drip irrigation: Install drip system with main and sub-main pipes and place the inline lateral tubes at an interval of 1.5 m.
- Place the drippers in lateral tubes at an interval of 60 cm and 50 cm spacing with 4 LPH and 3.5 LPH capacities respectively.
- Sowing: Dig pits of size 45 x 45 x 45 cm at spacing of 2 m in row spaced at 1.5m. Sow three seeds in each pit.
- After germination, thin to 2 healthy seedlings.
- Instead of direct sowing, the seeds can be sown in poly bags @ 2 seeds/ bag and 15 days after germination, they can be planted @ 2 seedlings per pit.
- Fertigation: Apply a dose of 250: 100: 100 kg NPK/ha throughout the cropping period through split application.
- Fertigation Schedule- Ribbed gourd (Hybrid) Recommended dose: 250: 100: 100 Kg/ha
- After cultivation: Hoeing and weeding thrice and provide support for the plants to reach the pandal erected at a height of 2 m.
- Spray Ethrel 250 ppm (2.5 ml/10 lit. of water) four times commencing from 15th day of sowing at weekly interval to increase yield.

- Beetles, fruit flies and caterpillars: Spray Dichlorvos 76 per cent EC6.5 ml/10 lit. or Trichlorofon 50 per cent EC 1.0 ml/l. Donot use copper and sulphur dust, which are phytotoxic.
- Diseases: Powdery mildew: Spray Dinocap 1 ml/l. or Carbendazim 0.5 g/l.
- Downy mildew: Spray Mancozeb or Chlorothalonil 2 g/l. twice at 10 days interval.
- Yield: 14 – 15 t/ha in 125 days.

Sponge Gourd/Nenua/Ghewada; *Luffa cylondrica* (Smooth gourd)

- ☆ Cucurbitaceae 2n=26
- ☆ Origin-India
- ☆ Luffa: its name from the product loofah used in bathing sponge door-mats and also for cleaning utensils
- ☆ Gelatinous compound present in luffa (luffin)
- ☆ Higher protein found in this crop
- ☆ Sandy loam: Soil is best for its and soil pH= 6.0- 7.5
- ☆ Long warm season is best for its cultivation.
- ☆ Optimum temperature for its 25-27 ^{0}C
- ☆ Seed rate- 2.5- 3.5kg/ha spacing 1.5 -2.0 x 0.6- 1.2m
- ☆ Seed deep in water for 24hrs for rapidly germination
- ☆ Commercial training system= Kniffin system
- ☆ Bowar and Pandal system also used in this height (1.5-2.0 m)
- ☆ Sex ratio can be governed by exogenous application of ethrel @ 200-250ppm increase the female flower
- ☆ Crops ready for harvest= 60-90 DAS
- ☆ If delay harvesting fruit become more finer and un-easy feel for human consumption
- ☆ Fruits are smooth so called as ghiya and tori
- ☆ Flower colour deep yellow
- ☆ Anthesis time morning time
- ☆ Yield 80-120kg/ha
- ☆ Ancestor - ***Luffa graveolens***
- ☆ Seed colour: white
- ☆ Edible part= Immature green dioecious spp.- ***Luffa echinata***
- ☆ Scientific trend of evaluation in Luffa is given by Dutt and Roy
- ☆ **Variety; IARI New Delhi**
- ☆ Pusa Chikni - Spring summer and rainy season
- ☆ Pusa Sneha – Long transportation
- ☆ Pusa Supriya- Summer and rainy season improve over Pusa Chikini
- ☆ (IIVR) Kashi Divya- Tolerant to P.M.D. and anthracnose

- CSA Kanpur-Kalyanpur Hari Chikni for rainy season
- AAU-Aanand Torai Chikini-1 rainy season
- HARP Ranchi- Swarna Prabha- Tolerant to PM, DM and leaf curl
- MH- Phule Prajakta for summer and kharif season

Pointed Gourd (Parwal: King of gourd): *Trichosanthus dioica (Roxb.)* 2n = 22

- ☆ Origin - Indo-Malaya (India)
- ☆ 2 per cent protein, 0.03 per cent fat, 153 IU Vit. A
- ☆ It is a Dioecious crops
- ☆ Type of fruit fuziform
- ☆ 13-20 cm long perennial cucurbits crops
- ☆ 1 male flower sufficient = 6-8 female flower
- ☆ Parwal is a dioecious climber with perennial root stock
- ☆ Parwal remain dormant during winter
- ☆ Rainfall is considers very good for this crops
- ☆ Propagation Wine cutting root suckers, pH range 6-7
- ☆ About 2000-2500 cutting (1m length) cutting are required
- ☆ 10-12 per cent male flower polinizer for fruit set 21-35 ^{0}C
- ☆ Optimum temp. for growth and development 25-35 ^{0}C
- ☆ Mid November ideal time in diara land for sowing
- ☆ Feb- march is ideal for upland for sowing
- ☆ In river bed 3500-4000 cutting moist lump method also used
- ☆ Flat system 4500-5000 cutting (1.0-2.0m spacing)
- ☆ Fruit availability February to October
- ☆ Every 15 female plant = one male plant gives high yield
- ☆ Lunda lacchi methods (8 shaped) used in pointed gourd
- ☆ Bower system =8000 cutting (1.0 x 1.5 m)
- ☆ Yield: 6-8t/ha 1st year 14-15t/ha 2nd year; Seed yield 250-300 kg/ha
- ☆ Recalcitrant seed found
- ☆ Fruits for seed extraction are harvested when turn yellow in colour
- ☆ Rice straw mulch @ 15-t/ha. give high yield
- ☆ Good crops for river bed cultivation
- ☆ Primary centre of origin is Bengal –Assam area
- ☆ Salicylic acid enhanced post-harvest life of fruit
- ☆ Major pollinator of pointed gourd: red brown beetle *Carpophyllus dimidiatus*

- ✰ Dandli Guli, Bihar Sharif, Kalyani, Faizabad Parwal 1,3,4,5
- ✰ Damodar, Kazil, Bombay (W.B)
- ✰ CHES Hybrid-1 1st hybrid of pointed gourd
- ✰ Green Oval, Green Long Striped (M.P)
- ✰ Swarna Rekha-rainy season resist to (BW) and fruit-fly HARP Ranchi
- ✰ Rajendra Parwal-1 tolerant to long distance transportation
- ✰ Rajendra Parwal-2 vine and fruit rot tolerant to fruit fly
- ✰ Swarna Allaukik rainy season, better keeping quality, resist to (BW) and fruit-fly, sweet preparation
- ✰ HARP Ranchi
- ✰ Kashi Alankar= spindle shaped
- ✰ Narendra Parwal 260 tolerant to wilt complex
- ✰ Narendra Parwal 307 good self life
- ✰ Narendra Parwal 604 strip less light green colour
- ✰ Sonalika is the private var.
- ✰ Spacing 2x 2 m
- ✰ Ready to harvest 80-90 days after transplanting
- ✰ Saline and sodic soil is unsuitable for this crop
- ✰ Highest dietary fiber content 3gm fiber/100gm
- ✰ Colon problem due to very high dietary fiber content
- ✰ Fruit is particularly recommended for convalescence
- ✰ Kinetin (50ppm for fruit set)
- ✰ Pointed gourd is cross pollinated
- ✰ Anthesis take place during at night
- ✰ It is said to be good for maintaining healthy heart and brain
- ✰ Coastal climate is not suitable for this crop
- ✰ Male and female ratio for essential maintenance= 1: 10
- ✰ Germination of seed very poor
- ✰ Plant raised by root cutting start bearing within 10-12 weeks
- ✰ The yield of pointd gourd starts declining after third year
- ✰ Hot and humid climate best for its cultivation
- ✰ Pointed gourd is available in four different forms small, round dark green and long dark green

Ash Gourd: *Benincasa hispida* Cogn; F.- Cucurbitaceae 2n=24

- Hairy melon, Chinese preserving melon
- Monoecious in nature
- ***Benincasa* is monospecific**
- It is good for people suffer from (nervousness)
- **The Ayurvedic Kooshmanda Asayan prepared from ash gourd.**
- Highest amount of water found (96.5 per cent)
- Fruit weight 2-10 kg
- Chromosome number2n=24
- Seed rate 5-7kg/ha
- Optimum temp. 24-30°C
- Mudliar-
- Pusa Ujjal- 1st improved var. ideal for petha preparation
- Kashi Dhaval- highest suitable for prepation of petha
- CO-1, CO-2- early maturing var.
- Pusa Shakti
- Kashi Surbhi- distant hybrid
- yield-10-15tonnes/ha
- Susceptible to frost
- Growing time June-July for October- harvest
- River bed- Growing (Feb.-March)
- Longest self life among cucurbits
- Male female ratio- 25-30: 1 to 15: 1
- Short day promotes female flower production
- Primitive sex form of cucurbits- hermaphrodite
- Recent sex for- monoecious
- Varieties
- Pusa Ujjawal- 1st improved variety, highly flesh recovery, ideal for petha preparation of ash gourd.
- Kashi Dhamval -Highly suitable for preparation of petha.
- Kashi Shurbhi (Distant hybrid)

- ☆ Mudliar, Push Shakti
- ☆ Co1, Co2 (early mature variety)
- ☆ Soil: A deep loamy soil with pH range of 6.5-7.5 is suitable. Opt. Temp.24-30 ^{0}C
- ☆ Season and sowing July and January.
- ☆ River bed growing (Feb.- March)
- ☆ Longest self-life among the cucurbits
- ☆ Three seeds are sown in each pit.
- ☆ The seeds are treated with *Trichoderma viride* @ 4 g or *Pseudomonas fluorescens*@ 10g or carbendazim @ 2 g/kg of seeds and after germination, the seedlings are thinned to two per pit.
- ☆ Seed rate: 5-7 kg/ha.
- ☆ Yield: 10-15 t/ha.
- ☆ Susceptible to frost
- ☆ Soak the seeds in double the quantity of water for 30 minutes and incubate for 6 days.
- ☆ Preparation of field Plough the field 3–4 times.
- ☆ Dig pits of 30 cm x 30 cm x 30 cm at a spacing of 2 x 1.5 m and form basins.
- ☆ Irrigation: Irrigate the basins before dibbling the seeds and thereafter once a week.
- ☆ Fertigation Schedule- Recommended Dose: 200: 100: 100 kg/ha
- ☆ After cultivation: Hoeing and weeding thrice. Spray Ethrel 250 ppm (2.5 ml/10 lit. of water) four times at weekly intervals commencing from 15th day after sowing.
- ☆ Quality seedling production: use 12 days old healthy seedlings obtained from shade net houses for planting.
- ☆ Raise the seedlings in pro-trays having 98 cells. Use well decomposed coco-peat is used as medium. Sow one seed per cell.
- ☆ Water regularly twice a day.
- ☆ Transplant about 12 days old seedlings in the main field
- ☆ Fertigation: Apply 60: 30: 30 kg of NPK/ha throughout the cropping period through split application.
- ☆ 75 per cent of phosphorus, is applied through superphosphate as basal dose.
- ☆ Pests; Fruit fly: fish meal trap with 5 gm of wet fish meal and 1 g of di-chlorophos in cotton, 50 traps are required per hectare.

- Fish meal and dichlorphos impregnated cotton are to be renewed once in 20 and 7 days respectively. 5. Neem oil @ 3.0 per cent as foliar spray as need based.
- Aphid: Spray Imidachloprid @ 0.5 ml/lit. along with sufficient quantity of stickers like Teepol, triton X100.
- Powdery mildew: Spray Dinocap 1 ml/l or Carbendazim 0.5 gm/l.
- Downy mildew: Spray Mancozeb or Chlorothalonil 2 g/l twice at 10 days interval.
- Yield: 20 t/ha in 140 days.

Snake Gourd: *Trichosanthes cucumerina*. L. F.- Cucurbitaceae

- Sandy loam rich in organic matter with good drainage and a pH range of 6.5-7.5.
- Season and sowing: July and January.
- Sow the seeds (3 seeds/pit) treated with *Trichoderma viride* @ 4 g/kg or *Pseudomonas fluorescens* @ 10 g/kg thin the seedlings to two per pit after 15 days.
- Seed rate: 1.5 kg/ha.
- Preparation of field: Plough the field to fine tilth. Dig pits of size 30 cm x 30 cm x 30 cm at 2.5 m x 2 m spacing and form basins.
- Irrigation: Irrigate the basins before dibbling the seeds and thereafter once in a week.
- Application of fertilizers: Apply 10 kg of FYM, 100 g of NPK 6: 12: 12 mixture as basal dose per pit and N @ 10 g pit 30 days after sowing.
- Apply *Azospirillum* and Phosphobacteria @ 2 kg/ha and *Pseudomonas* 2.5 kg/ha along with FYM 50 kg and neem cake @ 100 kg before last ploughing.
- Hoeing and weeding thrice. Provide stakes for the plants to reach the pandal (2 m).
- Spray Ethrel 100 ppm (1 ml in 10 l of water) four times starting from 10 to 15 days after sowing at weekly intervals.
- Nursery raising: Sow the snake gourd seeds in pro-trays containing well decomposed coco-peat medium.
- Sow only one seed per cell. Keep the trays under shadenet house.
- Water regularly with the help of rose can. Transplant about 12 day's old seedlings to main field.
- Planting spread the lateral tubes on the raised beds of 120cm wide at 150cm spacing.
- Irrigate the beds by operating the drip system continuously for 8-12 hrs.
- Plant the seedlings in the holes made at 60cm spacing.
- Fertigation: Apply a dose of 75: 100: 100 kg NPK/ha throughout the cropping period through split application.
- Pests; Leaf beetles and leaf caterpillars: Spray Dichlorvos 76 per cent EC 6.5 ml/10 litre or Trichlorofon 50 per cent EC 1.0 ml/litre.

- Fruit fly: Use 20 x 15 cm poly bags fish meal traps with 5 g of fish meal + 1 ml of dichlorvos in cotton @ 50 traps/ha.
- For management of Aphid vector, spray Imidachloprid @ 0.5 ml/litre along with sufficient quantity of stickers like Teepol, triton X100, apsa *etc.*
- Do not use copper and sulphur dust because these are phytotoxic.
- Powdery mildew: Spray Dinocap 1 ml/l or Carbendazim 0.5 g/l.
- Downy mildew: Spray Mancozeb or Chlorothalonil 2 g/l twice at 10 days interval.
- Yield: 18 t/ha in 135 – 145 days.

Ivy Gourd; *Coccinia grandis*

- Chromosome number- 2n=24
- Common name- Tondli/Little gourd/Scarlet gourd
- Origin- South-India
- Vit. A- 8000-18000 IU/100gm
- Related species- ***Coccinia abyssinica***- Bear edible tubores root
- Bitterness in ivy gourd- cucrbitacin
- Used for covering sugar content in diabetic patient
- Edible part – Immature fruit
- Planting time june July, Feb- March
- Sulabha- small fruit 9.25cm, weight 18.48gm
- Neelanchal Kunki CHIG-15- Commercial propagation- Cutting
- Sequential fruiting habbit found in ivy gourd
- Optimum temp for growth and development 30-35 ^{0}C
- Coccinia- useful for preaparion against cure- Bronchitis/diabeties
- Plant starts fruiting- 10-12 week after planting
- Ivy gourd fruit are produce almost through out the year
- Average 200-300 fruit weight 3-4 kg/vine/year
- Under ideal condition 8-10 kg fruit/vine/year

Tinda: *Citrullus vulgaris* var. *fistulosus*; Schrad, F.- Cucurbitaceae

- Varieties: Arka Tinda.
- Soil: Sandy loam rich in organic matter with good drainage and pH range of 6.5-7.5.
- Season of sowing: January-February Sow the seeds on one side of the channel
- Seed and rate: 3.5 kg/ha. Treat the seeds with *Trichoderma viride* @ 4g or *Pseudomonas @ fluorescens* 10 g or carbendazim 2g/kg of seed.
- Thin the seedlings after 15 days to maintain two per pit at 0.9 m spacing.
- Preparation of field: Plough the field to fine tilth and form long channels at 1.5m apart.
- Irrigation: Irrigate the field before dibbling the seeds and thereafter once a week.
- Application of fertilizers: Apply FYM 10 t/ha, N 20 kg/ha as basal and N 20 kg/ha 30 days after sowing.
- Apply *Azospirillum* and Phosphobacteria @ 2 kg/ha and *Pseudomonas* @ 2.5 kg/ha along with FYM 50 kg and neem cake @ 100 kg before last ploughing.
- After cultivation: Hoeing and weeding thrice.
- Pest; Beetles: Spray Malathion 50 EC 1 ml/l at weekly intervals. Collect the affected fruits and destroy.
- The fly population is low in hot day conditions and it is peak in rainy season.
- Hence adjust the sowing time accordingly.
- Expose the pupae by ploughing.
- Use 20 x 15 cm polythene bags, fish meal trap with 5 gm of wet fish meal and 1 g of dichlorvos in cotton, 50 traps are required per hectare. Fish meal and dichlorvos impregnated cotton are to be renewed once in 20 and 7 days respectively.
- Neem oil @ 3.0 per cent as foliar spray as need based.Do not use copper and sulphur dust, as these are phytotoxic.
- White fly: Spray Neem Seed Kernal Extract 5 per cent. Yield: 10 t/ha in 90 days.

Chow Chow: *Sechium edule*; Cucurbitaceae

- Origin: South Meaxico and Central Asia 2n= 2x= 28
- Progenitor-***Sechium edule var. nigrum***
- Common name= Choco, Askas and Chayote
- Varieties: Green fruited and White fruited.
- **Single seeded vegetables**
- Fruit bearing season- October to November
- Requires well drained acidic soil with a pH of 5.5 - 6.5 and thrives best in a temperature range of 18 – 22°C and at an altitude of 1200 - 1500m.
- In plains, it comes up well during winter season.
- Edible part- Fruit and root
- Good source of Vit. C
- The tuberous part of the root is starchy.
- In India and Nepal, the plant and fruit is called **(Ishkus) and in China (Buddha palm)**
- Propagation by single seeded fruit
- Flowering start- 3-4 month after planting
- Male and female flower produce same leaf axil
- Fruit is roughly pear shaped
- **Viviparous germination habit**
- Dig pits of 45 cm x 45 cm x 45 cm at a spacing of 2.4 x 1.8 m.
- Fill up the pits with 10 kg of FYM, 250 g of urea, 500 g of super phosphate and 500 g of muriate of potash.
- Highly sensitive to frost
- Season and sowing: Hills: April - May. Fully matured and sprouted fruits collected from high yielding vines are planted in pits @ 2 –3 per pit.
- After cultivation: Hoeing and weeding are done as and when necessary.
- At initiation of vine growth, stake the plants.
- Provide pandal at a height of 2m. Prune the plants to ground level during winter from second year after planting.
- In hills, pruning period is January. Apply 250 g of urea to each vine after pruning and at the time of flowering.

- ☆ Pests: For scales, mealy bugs and aphids, spray Dimethoate 30 EC @ 1 ml/l.
- ☆ Fruit fly: 50 traps are required/ha, fish meal + dichlorvos soaked cotton are to be renewed once in 20 and 7 days respectively.
- ☆ Yield: 25 - 35 tonnes/ha.

Chapter 4

Leafy Vegetables

Amaranthus: *Amaranthus* spp L.; Amaranthaceae

- Origin- India
- Chromosome number- 2n=32/34/64
- Type of fruit- Glomerule
- Poor man's vegetable
- Day neutral, C_4 plant
- Warm season crop
- Susceptible to water logging condition
- Seed rate- 1-1.5 kg/ha- for transplanted
- 2-2.5 kg/ha for direct sowing
- Seed yield-200kg/ha
- Leaves yield 100-150q/ha
- Harvesting -30 DAS
- Yield 7-19 tonnes/ha
- Isolation distance- 400m for seed production
- Mainly grown in rainy and summer season
- Grain type amaranthus is widely grown in Gujrat/MH
- Grain type- cross pollinated- wind pollination
- Vegetable type- self pollinated

- Variety- shallow rooted
- *A. cruentus*- Shortday
- *A. hypochondriacus*- Day netral
- *A. caudatus*-Ornamenral love lie bleeding
- *A. blitum*- Clipping type, 6 clipping can take 2week interval
- *A. dubicus*- Tetraploid spp., 2n=64
- Mainly grown in India- *A. tricolor*, 2n=34
- Protein contains 2-4 times more essential amino acid then normal amaranthus
- Major disease- White rust- *Albugo bilitiis*
- Seed depth-1.0-1.5cm
- Mostly cultivated monoecious
- Spacing 20-25cm Choti chauli
- 30-35cm Badi chauli
- Average spacing 30-30cm
- Main objective- low oxalic acid and high Ca contain
- Clipping is done in amaranthus
- Choti Chaulai-*A. blitum*
- Badi chaulai- *A.tricolor*- Commercial cultivation
- Pusa Kirti—Commercial cultivation
- *A. caudatus* is grown for seed purpose
- Grain amaranthus are highly resistant to drought
- South Indian green leaves spp.- ***A.mangostus/A.lividus/A.dubius***
- Sandy loam, salinity acidic is preferred
- Ph-6.4
- Loncelote var.- Shape-origin India
- Pusa Kiran- Broad/ovate/Domica(rainy season)
- Arka Arunima—Low anti-nutrient factore- oxalate, nitrogen
- Water requirement- high
- Seed treatment- succinic acid- increase yield without affecting quality
- Introduced tetraploid var.- CO5
- Acid- citric acid
- Arka Suguna- Resistant to white fly

- ✰ Pusa Kirti- Cultivation in summer season
- ✰ Pusa Lal Chaulai- *A.tricolour*
- ✰ 65-genera, 850 species,
- ✰ Soil: Well drained loamy soils with slightly acidic nature and warm climate are suitable.
- ✰ Season and sowing: Can be sown throughout the year.
- ✰ Broadcast the seeds evenly on the bed after mixing with 10 parts of sand.
- ✰ Preparation of field: The field is prepared to a fine tilth and beds of 2 m x 1.5 m are formed.
- ✰ After germination, thin the seedlings to have a spacing of 12 – 15 cm.
- ✰ Irrigation: Irrigate before and after sowing and at weekly intervals after germination.
- ✰ Application of fertilizers: Apply FYM 25 t/ha, *Azospirillum* 2 kg and Phosphobacteria 2 kg/ha, N 75 kg and K 25 kg per ha as basal dose.
- ✰ Pests;Ants: Apply lindane 1.3 per cent dust @ 10 kg/ha around the beds to control ants, termites and other burrowing insects.
- ✰ Leaf eating caterpillar: Carbaryl 50 WP @ 2 g/l.
- ✰ Leaf spot: Carbendazim @ 1 g/l of water. Spraying sulphur compounds should be avoided.
- ✰ Harvesting and yield: Leafy types 25 days after sowing for Mulaikeerai (10 t/ha); 40 days after sowing for Thandukeerai (16 t/ha).
- ✰ Clipping types: 10 clippings at weekly intervals (30 t/ha).

Curry Leaf; *Murraya koenigii* Linn.; F. Rutaceae

- Varieties: Sen kambu, Dharwad-1, Dharwad-2
- Red sandy loam with good drainage will be ideal for its normal and fleshy growth, which will result in better leaf yield.
- The optimum temperature requirement is 26 to 37 ^{0}C.
- The main season of availability of curry leaf fruits is July- August within 3-4 days of collection of fruits,
- The seeds should be pulped and sown in nursery beds or poly bags.
- One year old seedlings are suitable for planting.
- One seedling is planted at the centre of thc pit.
- The field is ploughed 3-4 times to get a fine tilth.
- Before last ploughing, well decomposed FYM is applied @ 20 tonnes/ha. Pits of 30x30x30cm are dug one to two months before planting at a spacing of 1.2 to 1.5 m.
- Immediately after planting the pits are irrigated.
- On the third day the second irrigation is given, then the irrigation is given once a week.
- After each harvest, 20 kg of FYM/plant is applied and mixed with soil.
- Inter cultivation: Periodical hoeing has to be given. During first year, one intercrop like pulses can be taken.
- After attaining 1 m height, the terminal bud is cut off to encourage basal branching.
- In total, 5-6 branches are maintained per bush. Ten to twelve months after planting, the first harvesting starts.
- Citrus butterfly: Hand picking and destruction of the larvae.
- Leaf spot: Spray carbendazim @ 1 g/l of water. Spraying sulphur compounds should be avoided.
- Harvesting and yield: At the end of first year 250-400 kg of leaves/ha can be harvested.
- In II year: Once in 4 months, every time 1800 kg/ha, which would work out to 5100 kg/ha/year.
- III year: Yield 5400 kg/ha
- IV year: 2500 kg/ha once in 3 months, which would work out to 10,000 kg/ha/year
- 5th year onwards: 5000 kg/ha once in 3 months, which work out to 20,000 kg/ha/year.

Agethi: *Sesbania grandifolia*, F.- Leguminaceae; 2n=18

- Leaves are rich in Vitamin- A 25.56 IU/110 gm iodine 2mg/gm
- Flowers are also used in vegetable
- Tropical perennial tree valued for its leaves and flower
- Flower and leaf are having medicinal properties
- The barks contain tannin and yield good fiber and gum
- Average rainfall 1000-2000mm
- Temperature bellows 26-36 ^{0}C maximum and minimum 18-24 ^{0}C
- Black cotton soil is best for this
- Propagation by seeds
- It is resistant to drought
- Direct sowing spacing (90x90cm)
- Pit size 30x30cm.
- Rainy season is ideal for sowing time
- It can be used for standard for black paper and paan (betel-vine).
- A tree yield (4.5- 9kg of leaf/year)
- It can tolerate seasonal waterlogging
- Two types of flowers- Red flower and white flower (suitable for kitchen garden)

Lettuce; *Lactuca sativa*

- ☆ Origin- Mediterranean region
- ☆ Family- Compositeae
- ☆ Chromosome number- 2n=18
- ☆ Edible part- Tender leave and head
- ☆ Inflorescence- Achene
- ☆ Vit C 10mg/100g
- ☆ Self pollinated crop
- ☆ Proginatore- ***Lactuca serriola***
- ☆ Vit A 900IU/100gm
- ☆ Ca 50mg/100gm
- ☆ Very shallow rooted crop
- ☆ Major salad crop in North
- ☆ Sensitive to highly acidic soil
- ☆ Major salad crop in North America
- ☆ Chopping is done in lettuce
- ☆ Total morphological type 6- Crisp head/Butter head/Stem/Leaf/ Romaine/Lectine
- ☆ Amphidiploid in nature
- ☆ Rich source of lutien
- ☆ ***Lactuca serriola***- Resistant to insect, drought, β-carotene
- ☆ C3 plant
- ☆ Indian lattuce- *Lactuca indica*
- ☆ Wild type/prickly lettuce- ***Lactuca serriola***
- ☆ ***Lactuca virosa***- Resistant to DM- lettuce mosaic virus
- ☆ ***Lactuca serriola***- Thermo dormancy found in lettuce
- ☆ Celery and lettuce seed have- thermo dormancy
- ☆ Mulching is most essential operation in lettuce cultivation
- ☆ Russet spotting due to ethylene injury
- ☆ Lettuce seed germinate- 4-5 days
- ☆ Recommended CO_2 in lettuce cultivation 1000-1500ppm
- ☆ Chilling treatment given seed 4-6 ^{0}C for 3-5 days break dormancy

- Seed rate- 400-500gm/ha
- Yield- 10-12tonnes/ha
- Leaf type-500-600kg/ha
- Head type- 100-125kg/ha
- Temperature 13-16 ^{0}C for seed germination
- High temp. induces bolting also
- Well drained sandy loam rich in organic soil is best
- Natural soil or slightly acidic pH 6.0-6.5 best
- Var.- Punjab lettuce -1 – non headed
- Var.- Imperial-856
- Leaf type lettuce- Slobolt Chinese yellow
- Butter head- White baston/big baston
- Tinged brownish colour- Iceberg
- Stem type lettuce- Celtuce
- Great Lake – Wholly green colour
- Crips head cultivar and resistant to tip burn.
- Imperial 859, New York 515
- Butter head cultivar- Resikia wonder var. voorefing
- Long day plant
- Mallic acid, citric acid
- Mary king
- Cos or romainer- Loose shaped head- Dark green, Eiffel Tower
- Classification given by- Thomson and Kelly
- Pest and disease- lettuce mosaic virus- seed borne disease
- Aphid transmission
- Downey mildew- ***Bremia lactucae***- Most serious pest
- Botrytis rot- ***Botrytis ciner***a- serious disease in Green/Dark Green Eiffel Tower, Paris Island, Paris White (loose head type)
- Most imp. disease in head lattuce slime soft rot
- Post harvest disease- Botrytis rot/soft rot, ***Erwinia carotovora***
- High temperature above 22 ^{0}C accelerate seed stock
- **Physiological disorder**- Tip burn- Due to calcium chloride nutritional deficiency light maturity and Ca/Mn/Boron

- ☆ Excess nitrogen/high IAA/soil moisture ontogenic age of plant
- ☆ Big vein is in lettuce due to Ca and boron deficiency
- ☆ Rosette spitting is the disorder of lettuce due to ethylene injury
- ☆ Seed will not germinate at temperature above 30 ^{0}C
- ☆ Planting early in Oct-Nov 5-6 week after sowing
- ☆ Space 45-45 cm,30-45cm
- ☆ It is highly sensitive to highly acidic soil
- ☆ Storage 3-4 week under refrigerated condition
- ☆ Age of transplanting 4-5 weeks
- ☆ Transplanting early to transplanting
- ☆ Lime requirement very high
- ☆ Root depth very shallow 15-30cm tolerant to soil salinity
- ☆ Pre-cooling (-1 ^{0}C)

Celery; *Apium graveoleus*

- ☆ Origin- Italy
- ☆ Family- Apiaceae
- ☆ Toxic substance- Apin
- ☆ Edible part- Petiole
- ☆ It is used for flavouring veg, juices and soups
- ☆ Seeds are used as condiments
- ☆ Shallow rooted.
- ☆ 2-3 per cent volatile oil present in seed
- ☆ High temp cause bitterness in leaves
- ☆ It is biennial crop
- ☆ Crop Bot when temp fall below 15 °C
- ☆ 2nd importance amongst salad crops
- ☆ It is slighty sensitive to acidic soil
- ☆ pH 6.6 optimum
- ☆ ***3-N butylphthalide***- found in celery that effective against - effective against hypertension
- ☆ It is two type-self branched or yellow
- ☆ Green leaves
- ☆ Green leaves var. preferred in the Indian market
- ☆ Florida Golden and Golden belong to self branched
- ☆ White Grove Giant and Fordhook Emperor and Standard Bearer
- ☆ Green leaves pod
- ☆ Bolting initiation when temp. below 15 °C
- ☆ Emperor is a var. of celery
- ☆ Propagation by seed
- ☆ 100-125gm seed/ha
- ☆ Malic acid present
- ☆ Toxic substance apiin
- ☆ Seed germination is good when temp is 15-20 °C
- ☆ Spacing 60cm×20cm
- ☆ ***3-N butylphthalide*** is very effective against hypertension

- ☆ Seed shattering is major problem
- ☆ Optimum temperature 13-24 ^{0}C
- ☆ Tender leaves harvested in May month in plain
- ☆ Seed rate of celery 250gm/ha
- ☆ Yield 330-560q/ha
- ☆ Above 24 ^{0}C 'so stalk becomes fibre and bitter flavour
- ☆ Inflorescence-Umbel

Parsley; *Petroselinum hortense*

- Origin- Mediterranean region
- Chromosome number- 2n=18, 22
- Edible part- leafy stalk and petioles
- Rich source of iron vit-A and vit-C
- It is used for garnishing and flavouring
- It can be produces under greenhouse and polyhouse condition
- It is biennial herbaceous plant
- Seed rate: 25-30 kg/ha sufficient
- Seed are sown Feb.- March, Hill area August - Oct.
- Its take 18-24days to germinate
- 12-20 cm plant to plant spacing
- 30-45 cm row to row spacing
- Two months old seedling become ready for transplanting
- Transplanting spacing 60cm×45cm
- Hight of parsley 30 cm-35 cm
- **Three type of variety-**
- Plained leaved curls, evergreen
- Double curled
- Moss curled- Curl leaf commonly knows as curl leave parsley var.crispon
- Moss curled- Its is used for garnishing and decorative
- Hamburg- Tall var./rich in vit-C/90mg/100gm
- Irrigations are given 15-20 days intervals
- Neapolitan- celary leaved parsley
- Respiration rate- High
- Insect- aphid- phosdrin-1-2 per cent 5days interval before harvesting
- Carrot weevil- Guthian-2 per cent solution 21 days before harvesting
- 1st year only (leaf harvest)
- 2nd year only (seed harvest)

Spinach Beet/Palak; *Beta vulgaris* var. *bengalensis*

- ☆ Chromosome number- 2n=18
- ☆ Origin- Indo-Chinese
- ☆ Long day plant
- ☆ Herbaceous annual and binneal for seed production
- ☆ Rich source of Vit. A 9770 IU, Vit. C 70mg/100gm, Ca 380mg/100gm
- ☆ Palak is a tolerant to saline soil
- ☆ Ancestor of palak-*beta-vulgaris*. var. *maritma*
- ☆ Palak is closely related to beet root/sugar beet
- ☆ Beta vulgaris var. cicla and swiss chord
- ☆ Palak leaves contain low oxalicswiss chord
- ☆ Spinach beet is primarily used as pot herb
- ☆ Palak produce hermaphrodite flower
- ☆ Cross pollination due to wind pollination
- ☆ Single fruit of palak contain 2-3 seeds
- ☆ Palak seed germinate after 8-10 days after sowing
- ☆ Seed viability 3-4year
- ☆ Fruit is seed ball or multigerm 2-3 seed/fruit
- ☆ Average leaf cutting 4-6
- ☆ Average yield-600kg/ha
- ☆ It can tolerate frost better than other vegetable
- ☆ Sandy loam soil is most suitable
- ☆ pH 7.0
- ☆ Sowing of time-
- ☆ Rain- last week of august to 2nd week of Nov.
- ☆ Hill- March- April
- ☆ Seed rate- winter 10-15kg/ha
- ☆ Summer 25-30kg/ha
- ☆ Harvesting- 3-4 after sowing
- ☆ Flower- Bisexual flower pistilate
- ☆ Polyploid var.- Pusa jyoti
- ☆ Mutant var.- Jobner Green

- Root composite- Arka Anupama- It can tolerate to cercospora leaf spot
- Var.- Punjab Green – Mid purple pigment in stem
- Pusa Jyoti- Polyploid var.
- Pusa Bharati- Highly Vit. E and Beta carotene
- All Green- Selection local var.
- Pusa Palak- Swiss chard×Local palak
- Banargee jaint- Palak×Beet root
- Pusa Harit- Sugar beet×Desi palak

Bathua Wild Spinach; *Chenopodium album* L.

- Pigweed lambsquarters, fat hen (used for fattening poultry) goosefoot white
- South America
- Family: Chenopodiaceae (Amaranthaceae),
- They are now included under the family (amaranthaceae)
- It is one of the best sources of beta carotene, calcium, potassium, iron in the world
- Great source of minerals B-complex vitamins, Vit-C and fibre
- Plants are clumping nature with (2.0- 3.0m long, annuals)
- The leaf is diamond shaped
- It is a dicotyledonous nature
- Red variety is grown in Bolivia and black is grown in colardo (USA)
- It needs rainsing to remove any traces of bitter coating called- Saponin
- Quinoa is higher in protein than most of the grain
- The panicles have a central axis from which a secondary axis emerges either with flower (amaranthiform)
- Bearing a tertiary axis carrying the flower (glomeruliform).
- It is a pseudocereal rather than a true cereal.
- Blood purifying properties found in pigweed.
- It is propagated through seeds
- Seeds are slightly smaller than the mustard seed.
- Thirty days old seedling are transplanted in the main-field at a spacing 30cm x 15cm

Chapter 5

Bulb Vegetables

Onion - Small Onion (Aggregatum); *Allium cepa* var. *aggregatum*

- Botanical name- ***Allium cepa***
- Family- Alliaceae
- Sub falimy- Alliadeae
- Inflorescence- Umbel
- It is highly cross poliinated due to protandry
- India ranks 2nd in area and production in the world after China
- 3rd rank after Netherlands/Spain/India
- Maharastra is leading state 23.4 per cent are 27 per cent production
- Average yield/ha highest in Gujarat 23.37tonne/ha
- Per capita availability of onion is highest in Netherlands
- Pusa White Flat - TSS 11-12 per cent, drying ration 9: 1
- Pusa White Round – TSS 11-13 per cent, drying ration 8: 1
- Nitherland is the leading export of onion contributing to about 21 per cent of world export
- *Allium fistulasum*- multi disease resistance
- Green onion also used pickling
- Downey mildew resistance- *Allium praemitum, A. odorata, A. royeli*
- Raw onion are consumed as young green plant

- ☆ Some time called spring onion or as bulbs
- ☆ Mild flavoured or colourfull bulb used for salad
- ☆ In India prefer red and pungent type onion
- ☆ Europe and America yellow coloured preferred
- ☆ Allums are sulpher containing compound
- ☆ Onion oil yield methyl and propyl compound 0.002-0.003mg/100gm
- ☆ Fresh bulb
- ☆ Dry matter constant of allum- 7-15 per cent
- ☆ Bolting- initiating of seed stock
- ☆ Onion account for 77 per cent total foreign exchange earning among fresh veg.
- ☆ Test weight- 3gm
- ☆ Common onion- *Allium cepa* 2n=16
- ☆ Shallow- *Allium cepa acalanicum,* perennial onion
- ☆ Potato onion/multiper onion- *Allium cepa var. agrigatum*
- ☆ Tree onion/egytian tree onion- *Allium cepa var. viviparum/proliferrum* 2n=2x=16
- ☆ Potato onion/underground onion and multipher onion
- ☆ It is also known as egyptium ground onion
- ☆ It does not produce-any seed but remain in the ground as perennial lumph of bulb year after year
- ☆ Tree onion- It is viviparous plant that grow as perennial bulb
- ☆ It is resistance to pest and disease
- ☆ No of seed stem- 1-20
- ☆ No. of flower umbel-50-2000
- ☆ White coloured var.- Pusa White Flat- TSS 11-12 per cent
- ☆ Pusa White Ground TSS 12-13 per cent
- ☆ Yellow coloured var.- Early Grano/Brown Spanish- Arka Pitamber
- ☆ Male sterile lines have been isolated in Pusa Red at IARI
- ☆ Selfing and massing is suggested by Jones and Mann 1963
- ☆ Inbreeding depression is very common
- ☆ Onion can also be grown as intercrop in sugarcane field
- ☆ Good vegetative growth 12.8-23 per cent
- ☆ Bulb development 20-25 ^{0}C

- Brown Spanish- Long day variety, day length-> 14hrs
- Long day var.- Do not under, short day condition
- Where short day var. if planted under long day condition will developed early bulb
- Temp. is more important for seed production
- Photoperiod is more important- bulb formation
- Best time for sowing early kharif crop April-may
- For kharif season (May-June) in south
- Optimum time of transplanting, rabi season onion, north India mid Dec to mid Jan
- A set- 1-1.2 tonnes are bulblets are required for planting 1 ha
- Most favorable temp. for purple blotch 28-30C/RH 80-90 per cent in north India. It is very serious in seed crop
- Long day suitable for hill
- Mr. Onion nick name Dr. HA jones
- Ancestor of onion -*Allium vavillovi*
- Kurrat-*Allium kurrat*
- Ancestor of leek and kurrat- *Allium ampeloprasum*
- Onion seed viability one year
- Allium spp- Recurrent apomixis
- Red onions are more pungent than white onion
- Onion black colour seed controlled by single dominant gene
- CGMS- first found in onion
- Best quality onion seed- bulb seed method
- High yield of onion seed to seed
- Optimum soil ph 5.8-6.5
- NRC on onion and garlic- Nashik 1194 MH
- Yellow colour outer screen- quercetine
- Bulbing in onion is favored by- long day
- Edible portion- modified stem bulb
- Tear inducing action of onion 1propenl suyonic avid
- Onion richest source of vanadium
- New source of male sterility *Allium glaucum*
- Onion bulb culture is governed by major genes (ICAR)

- ☆ Onion bulb shape governed by multiple gene
- ☆ Onion downy mildew is controlled by single recessive gene
- ☆ New var.- Palam Lohit
- ☆ IARI Pusa Riddhi/high antioxidant variety
- ☆ Pusa Saumya- 1st blanching var.
- ☆ Storage disease- *Aspergillus penicillium*
- ☆ IIHR congumation maxican var. short day
- ☆ Arka Vishesh rabi season and suitable for export
- ☆ Arka Ujjawal
- ☆ Arka Pitamber- kharif and rabi
- ☆ Arka Sona- Rabi season export purpose
- ☆ Arka Bheem- Arka Akshay- Synthetic var.
- ☆ Arka Lalima and Arka Kirtiman- F1 hybrid developed by CMS line
- ☆ Bhima Raj, Shima, Red, B. Kiran, B. Shakti, Bhima Shweta
- ☆ Natural triploid var.- Pran
- ☆ 1st CMS identified in Italian seed by- HH jones
- ☆ Seed yield 800-850kg/ha
- ☆ Good keeping quality Pusa Madhuri/Agrifound Light Red
- ☆ Arka Akshay- Triparental and synthetic var.
- ☆ Long day onion- Brown Spanish
- ☆ Aster yellow- Viral disease
- ☆ Yellow dwarf disease- Viral disease
- ☆ Onion shoot is the modification of leaf sheath
- ☆ Drying ratio 6: 1
- ☆ White most of var. are dried at- 10: 1
- ☆ Anti-fungal factor in phenolics of onion- Catechol
- ☆ Hybrid seed production female/male ratio-4: 1
- ☆ Onion carbohydrate- Fructone
- ☆ Direct sowing Agri Found Rose- 22.5kg/ha
- ☆ Row to plant spacing
- ☆ 15-10 cm big onion
- ☆ 8-5 cm small onion
- ☆ Best time of planting- for kharif- July- August

- In north India mid or end of August
- North eastern India- December- January
- Harvesting (Dec.- Jan.)
- The best time of harvesting for rabi season onion is one week after 50-70 per cent neck fall
- Hissar 2, Punjab selection- better for salt tolerant
- Export purpose var.- Arka Bindu, Agri Found Rose, Banglore Rose
- Onion var. best for export purpose are- Red coloured and small bulb
- Harvesting 75-80 days after transplanting
- Dehydration pupose- Pusa Flate White, Punjab 48
- Onion recommended for growing as green onion
- Early grown- Pusa White Flate, Pusa White Round
- MDU 1, CO (On) 5 is a free flowering and seed setting type.
- Soil: Red loam to black soils with good drainage facilities.
- The germination and bulb maturation is affected in clay soil.
- It grows well in pH range of 6- 7 and a mild season without extreme of heat and cold.
- Season and sowing: Sow the medium sized bulbs during April-May and October-November.
- It requires sufficient soil moisture during its growing period but heavy rains during bulb germination and bulb formation affects the crop growth.
- Seed rate: Seed bulbs 1000 kg/ha. Medium sized bulbs are to be chosen for planting. Seeds @ 8kg/ha.
- Raising seedlings and transplanting: This is the most common method practiced for irrigated crop as it results in high yield and large size bulbs.
- In plains, seeds are sown during October-November for a Rabi crop.In hills, seeds are sown from March to June.
- Seeds are first sown in well prepared nursery beds of 90-120 cm width, 7.5-10.0 cm height and convenient length.
- Ratio between nursery area and main field is about 1: 20.
- Seed rate varies from 8 to 10 kg/ha.
- Seedlings of 15 cm height and 0.8 cm neck diameter are ideal for transplanting and this is achieved in 8 weeks.
- However, it varies from 6-10 weeks depending on soil, climate and receipt of rain.

- There is a practice of topping seedlings at the time of transplanting if seedlings are over-grown.
- Preparation of field: Plough the land four times to a fine tilth.
- Form ridges and furrows at 45 cm spacing.
- Sow the bulbs or plant the seedlings on both the sides of the ridges at 10 cm apart.
- Irrigation: Irrigate at the time of planting of seedlings and third day and later at weekly intervals. Withhold irrigation 10 days before harvest.
- Application of fertilizers: Apply FYM 25 t/ha, Azospirillum 2 kg and Phosphobacteria 2 kg/ha, N 30 kg, P 60 kg and K 30 kg/ha as basal and 30 kg N/ha on 30th day of sowing.
- Fertigation: Apply a dose of 75 per cent of the total recommended dose of superphosphate i.e. 285 kg/ha as basal dose.
- Raised beds of 120 cm are formed at an interval of 30 cm and the laterals are placed at the centre of each bed.
- Planting design: Bulbs or seedlings are planted in rows of 20 cm spacing and 12 cm between plants.
- Planting is done at 6 rows in each bed, thereby it accommodates 55560 plants in one hectare.
- Complete wetting is necessary to maintain uniform moisture level.
- Pests; Thrips and onion fly: Spray any one of the following insecticides based on the ETL of 1 maggot/hill. Insecticide dose Dimethoate 30 per cent EC 7.0 ml/10 lit. Oxydemeton –Methyl 25 per cent EC 1.2 ml/lit. Quinalphos 25 per cent EC 1.2 ml/lit.
- Cutworm: Drench the soil with chlorpyriphos @ 2 ml/l.
- Diseases; Leaf spot: Spray Mancozeb 2 g/l or Copper oxychloride 2.5 g/l. Add Teepol 0.5 ml/l of spray fluid.
- Basal rot: Seed or bulb treatment with *Trichoderma viride* @ 4g/kg and basal application of *T. viride* @ 2.5kg/ha along with VAM 12.5 kg/ha Selection of healthy onion seed bulbs.
- Bulb treatment–*Pseudomonas fluorescens* (5 g/kg) + *Trichoderma viride* (5 g/kg)
- Soil application of P. *fluorescens* (1.25 Kg/ha) + *T. viride* (1.25 kg/ha) + AM Fungi (VAM) (12.5 kg/ha) + Azophos (4kg/ha) + Neem cake 250 kg/ha.
- Installation of Yellow sticky traps 12/ha for thrips and leaf miner trapping.
- Installation of Pheromone traps 12/ha for cut wrom (*S .litura*).

- Spray application of *P.fluorescens* (5 g/l) + *Beauveria bassiana* (10 g/l) on 30DAP.Spray application of Azadairachtin 1 per cent (2 ml/l) on 40DAP.
- Need based application of Profenophos (2 ml/l) or Dimethoate (2 ml/l) or Triazophos (2 ml/l) for thrips/leaf miner/cut worm management.
- Need based application of Tebuconazole (1.5) ml/l) or Mancozeb (2 g/l)/ Zineb (2g/l) for purple blotch disease management.
- Harvest and Yield Spray Cycocel @ 200ppm + carbendazim @ 1000 ppm 30 days before harvest to extend the shelf life of onion.
- Harvesting is done by pulling out plants when tops are drooping but still green.
- During hot days when soil is hard, bulbs are pulled out with a hand-hoe.12-16 t/ha in 70 to 90 days. 18 t/ha in 90 days for CO (On) 5 onion.
- Clean and dry the bulbs for 4 days in shade soon after harvest.
- 5-Grade Specification: Shape, Size, skin colour, moisture content

Big Onion or Common Onion: *Allium cepa* var. *cepa*; Family- Alliaceae

- Soil: Red loam to black soils with good drainage facilities.
- The optimum pH is 5.8-6.5.
- Season: May – June. Mild season is preferred.
- Seed rate and Sowing: 10 kg/ha.
- Varieties: Bellary Red, Pusa Red, NP 53, Arka Niketan, Arka Kalyan, Agri Found Light Red, Agri Found Dark Red and Rose onion (small).
- Treat the seeds with *Azospirillum* @ 400 g/kg of seed using rice gruel as adhesive, dry under shade for 30 minutes and sown them.
- Apply VAM 1 kg/sq. m in the beds along with FYM 10 kg/sq.m before sowing.
- Preparation of main field: Plough the land to a fine tilth and incorporate 25 t/ha of FYM at the time of last ploughing.
- Irrigation: Irrigate at planting and third day and later at weekly intervals.
- Application of fertilizers: Apply FYM 25 t/ha, *Azospirillum* 2 kg and Phosphobacteria 2 kg/ha, N 50 kg, P 150 kg and K 75 kg/ha as basal dose and N 50 kg/ha as topdressing 30 days after planting.
- Pests; Thrips and onion fly: Spray methyl demeton 25 EC @ 1 ml/lit.or dimethoate 30 EC @ 1 ml/lit. with Teepol 0.5 ml/lit.
- Cut-worm: Drench the soil with chlorpyriphos @ 2 ml/l.
- Nematode: Application of Carbofuran 3 G or Phorate 10G @ 1 kg a.i./ha at 10 days after transplanting.
- Disease; Leaf spot: Spray Mancozeb 2 g/l or Copper oxychloride 2 g/l. Add 1 ml of Teepol to per ltr of spray fluid.
- Yield: 15 – 18 t/ha in 140 – 150 days.
- Grade Specification: Extra-large onion (>6 cm dia.), Medium (4-6 cm dia.), Small (2-4 cm dia.)
- Big onion and aggregatum onion production through fertigation
- Quality seedling production: The seedlings of big/common bellary onion are produced in raised nursery beds.
- Nursery area of 12.5 cents with slanting slope of 2 per cent is required for the seedling production to cover 1 ha.

- Unlike other transplanted vegetables, the nursery area of open/protected – raised beds may be used since, the seeds are too small and the number of seedlings requirement is also very high.
- Raised beds of 30cm height and convenient length at an interval of 30cm between beds must be formed for sowing the seeds.
- The beds are inoculated with *Arbuscular mycorrhizae* @ 1 kg/sq. m. as Seed treatment.
- Seed rate: 8-12kg of onion seeds are required to raise seedlings for one hectare of land.
- The seeds are treated with *Pseudomonas fluorescens* @ 10g/kg of seeds.
- Sowing: The seeds of onion are small, and hence the seeds should be mixed with fine sand and sown in lines 10 cm apart.
- After sowing, the seeds are covered by thin layer of sand and then by paddy straw.
- The nursery is watered through rose can twice a day regularly.
- On tenth day, copper oxychloride @ 2g/litre of water is sprayed to protect the seedlings from damping off disease.
- Field preparation: The field selected for planting should be ploughed thoroughly by using chisel plough, disc plough and cultivator.
- Along with biofetilizers 50kg each of zinc suphate and ferrous sulphate are applied before last ploughing.
- Raised beds of 1.2 m width and 30 cm height are formed for transplanting.
- Drip irrigation is done @ 8-12 hours depending upon the soil condition to get field capacity.
- Planting: Transplanting is done in rows at the spacing of 15 cm between rows (of 7 rows within the bed) and 10 cm between plants. A total population of 4,70,000 seedlings are required to cover one hectare of land.
- After care: Drip irrigation is done @ 1-2 hours per day depending upon the soil and climatic condition.
- Weeding: First weeding is done on 30thday after transplanting. Then weeding is done as and when necessary.
- Fertigation: For big onion, 60: 60: 30kg of NPK/ha is applied throughout the cropping period. The splits are approximately once in every 3 days.
- Harvesting and post-harvest management: The harvesting is done 75 to 160 days depending upon the Varieties.

- ☆ The irrigation is stopped 15 days before harvesting and sprayed with 2500 ppm of maleic hydrazide. This Swill prevent sprouting of the bulbs in storage, by which they can be stored even for 6-7 months. The correct time of harvest is one week after 50 per cent top fall.
- ☆ After lifting, the bulbs with tops should be cured in shade for 10-15 days before storage to remove field heat.

Garlic (*Allium sativum* L.) Family-Alliaceae

- Chromosome No. 2n=16
- Common name- Necter of life
- Egypt 1st in productivity
- Centre of origin - Central Asia
- Ancestor of garlic –*Allium longicuspis* pungency- Diallyl disulphide (60 per cent) diallyl trisulphide (20 per cent)
- Non-bolting spp. *Allium sativum var. sativum* and *Allium sativum var. vulgare*
- Bolting spp. *Allium sativum var. scordoprasum* is triploid 2n=3x=21
- *Allium sativum var. ophioscorodon*
- *Allium sativum var. sagitatum*
- Botanical Name of great headed garlic- *Allium Amploprassum*
- Amino acid found in injured garlic allin
- Enzyme responsible for breakdown of allin into allicin
- It is very hardy crops
- Garlic is produced only in one season rabi season
- Harvesting 4-5 month after transplanting.
- Delay harvesting bulb splitting.
- It reduces the cholesterol in the blood.
- Multicidal properties found in garlic (Nematicidal, Fungicidal, Bactericidal)
- Suitable range of pH is between 6-7
- Short day types are cultivated
- Critical day length for bulbing is 12hrs
- 2nd most important bulb crops in India.
- Yield: 10-20t/ha.
- Curing is done to remove the excess moisture., 7-8 days in shade.
- Bulb are stored at 0 °C -2.2 °C temp. and 60-70 per cent RH
- Essential oil present 0.1 per cent to 0.4 per cent sexually sterile diploid nature
- Garlic propagated by clove
- Bulbil is also propagated in some varieties. Tissue culture is used to produce healthy seedling

- China is the leading producer of garlic in the world.
- Anti-bacterial substance (allicin)
- For exporting purpose bulb size- 40-60 mm diameter 10-15 clove
- True garlic odour is due to presence of diallyl disulphide
- Earliest maturing var.- (G- 282)
- Garlic bulb (multiple bulb)
- Critical day length for bulbing (12hrs)
- Optimum sowing time (Oct-Nov.) harvesting Feb-March perha. Seed rate 5-6q/ha.
- Optimum spacing (10x 7.5cm), (15 x 10cm)
- Recovery percentage of clove from the bulb (86-96 per cent)
- Ascorbic acid is very high in green garlic leaves
- Garlic goods for cardiovascular activity
- Largest bulb size- Jamnagar Local
- In some cases garlic propagated by (inflorescences bulbil)
- Storage life prolonged mallic hydrazide @ 3000ppm at 3 weeks
- Agrifound White (G-41) TSS- 41 per cent,1st garlic released by CVRC Susceptible to purple blotch.
- Yamuna Safed (G-1) TSS- 38-40 per cent, thrips tolerant var. tolerant to purple blotch,cultivated all over India
- Yamuna Safed -2 (G-50), Total Sugar 40-41 per cent
- In small bulb= 20-30 small cloves
- Agri Found Parvati (G-313) Bigger size bulb (5-6cm diameter) T.S.S 38-42 per cent, export long day,hard neck type variety
- G-282 Bigger size bulb (5-6cm diameter) Total Soluble Solids 38-42 per cent, early variety
- (G-282, 313, Ladwa) - Export purpose variety
- Fawari and Rajalle Gaddi grown in Bellary dist
- Pink bulb-Godaveri, short day type
- Shweta – Short day type
- Bhima Onkar and Bhima Purple
- Suitable for hill –(Agri Found Parvati)
- Seed rate 500- 800kg/ha.
- Delayed harvest (splitting of bulb)

- Post harvest losses about 15-50 per cent depend upon variety
- Heavy soil for misshape bulb
- Productivity highest Jammu (13.9t/ha.), area and production (M.P)

LEEK; *Allium ampeloprasum* var. *porrum* L Syn. *Allium porrum* L

- 2n=2x=32 family-Alliaceae
- Leek is a non bulb forming tetraploid plant of onion family.
- Grown for its blanched stem and leaves.
- It is eaten raw alone or mixed in salads and cooked as flavouring in soups and stews.
- It is not grown in India on commercial scale but is a favourite vegetable in a kitchen garden.
- The climatic and soil requirements are the same as those for onion.
- Except large size it looks like green onion.
- Its cultivation is not commercial in India and grown only in kitchen garden.
- Leek is cultivated for blanched stems and leaves.
- Allium species are not evenly distributed within the northern hemisphere,
- Leek is thought to have been derived from the sand leek (*A. ameloprassum*) which grows wild in southern Europe, North Africa and the Middle East.
- Energy (kcal)77, Vitamin-30IU, Moisture (g) 78.9, Protein (g)1.8, Calcium (mg) 50, Iron (mg) 2.3
- Leek grow best in a cool to moderate climate.
- The Willamette Valley and Oregon Coastare ideal.
- It can be grown all the year round.
- A well aerated mineral soils but in heavy clay soils with both good drainage and good moisture retention capacity with a pH of 6.5-7 is best.
- Deep ploughing is recommended.
- Main varieties (August to October):
- American Flag, Jolant, Kilima, King Richard, Primor.
- Latefall-winter (October to December):
- Derrrick, Electra, Goldina, Goliath, Kilima, Tivi, Wintereuzen.
- Over winter (spring harvest):
- Carina, Conqueror (moderate bulbing),Eskimo and Siberia.
- These varieties mature in 80-120days.
- PPL-1 selection from exotic germplasm. Whole plant consumed.
- Leaves light green, swollen stem and do not form bulb.

- Ready in 150-160 days, good replacement for green onion.
- Palam Paushtik: alternative of green onion, for salad, soup and cooking.
- Matures- 140-150days yield of 300-350q/ha.
- August–October is the ideal time of sowing in plains and seedlings are planted when they are 15cm height.
- In hills, it is sown in March-April.
- Leek withstands heat and cold better than onions.
- Temperature is more important than day length in seed stalk development.
- It thrives well in higher altitude.
- Seeds are produced in India at higher altitude.
- About 5-7kg of leek seeds is required to be sown per hectare, but leeks are not commonly direct seeded.
- Leek seed, like other Alliums has very limited useful viability (less than 2years) unless stored under ideal conditions.
- Direct field seeding is possible but not recommended due to the lack of registered herbicides and length of time needed to harvest from direct-seeded plantings (8-12 months or longer).
- Leeks are normally transplanted.
- They should be seeded under protection early in the season (February to March) or in outdoor seedbeds (April to May).
- Leeks for harvesting in August-September should be sown from mid-February to mid-March.
- For harvesting in early winter sow the crop from mid-March to mid-April.
- For spring harvest, seed may be sown from April to mid-May.
- The right varieties (above) must be used for each harvest season.
- At transplanting, leek seedlings should be placed in furrows15cm deep.
- Furrows are filled in during cultivation and as the plants develop, the rows should be hilled up.
- Leeks should have a white shaft that reaches almost to the base of the leaves.
- Another method of transplanting is to drop each seedling into an individual hole, 15 cm deep, pressed into the ground.
- This procedure eliminates the need for filling of furrows but ridging should still be done.
- Seedlings are spaced at a spacing of 40x10cm.

- ☆ As oil test is necessary to determine phosphate and potash requirement.
- ☆ Apply 200-250q well rotten Farm Yard Manure or compost along with 80-100kg N, 60 kg P_2O_5 and 80 kg k_2O/ha.
- ☆ Ethephon @ 50mg/l as foliar spray should be applied 20-25days after sowing.
- ☆ NAA @ 50mg/l should be applied at 60 and 90 days after planting.
- ☆ A total of 12-15 inches of water may be required depending on planting date, seasonal variation and variety.
- ☆ Blanching is important in leek cultivation.
- ☆ It is done by covering the plants to a certain height so as to bleach them, which improves the quality of the crop.
- ☆ Leeks do not bulb or go dormant in the fall but continue to grow slowly.
- ☆ Small leeks can be sold starting in early August and varieties that have frost tolerance may be harvested throughout the fall and winter months.
- ☆ Single or multiple row harvesters can be used effectively.
- ☆ Leek yields approximately 925q green bulbs per hectare.
- ☆ Store leek seed at 0 ^{0}C and 95to 100percent relative humidity.
- ☆ Leeks should be kept satisfactorily for 2 to 3 months at 0 ^{0}C.
- ☆ Leeks should be cooled promptly after harvest to near 0 ^{0}C by hydro cooling.
- ☆ crushed ice or vacuum cooling and they should be kept at this temperature with high relative humidity throughout storage.
- ☆ Yellowing and decay develop rapidly at warmer storage temperatures.
- ☆ High relative humidity is essential to prevent wilting.
- ☆ The use of polyethylene film crate liners and of crushed ice can aid in preventing moisture loss.
- ☆ Good refrigeration will retard the elongation and curvature that develops in leeks at 10 ^{0}C or 21.1 ^{0}C.
- ☆ Respiration or heat evolution of leeks is about eight times faster at 21.1 ^{0}C than at 0 ^{0}C.
- ☆ Storage for 4-5 months at 0 ^{0}C is possible by using a controlled atmosphere (CA),
- ☆ The best CA contains from 1-3 per cent oxygen and from 5-10 per cent carbon dioxide.
- ☆ Atmospheres containing 15-20 per cent carbon di-oxide cause tissue injury.

- Leeks are commonly trimmed to12 inch length, bunched depending on diameter and often placed in polyethylene film bags.
- They are usually packaged in 5kg cartons or wire bound crates, holding 10 film bags, each 500 g
- Other crates may be packaged with 18-24 bunches with a net weight upto 15kg
- Onion maggot: (*Hylemya antique*)
- It is similar in appearance to a house-fly, though smaller.
- Spray malathion (0.05 per cent) or deltamethrin (0.0025 per cent)
- Onion Thrips: (*Thrips tabaci*)
- Spring- summer crop is affected more by thrips.
- Grow resistant varieties like Nasik Red and Spanish White.
- Apply malathion (0.05 per cent) orphorate10 G (1 kg/ha).
- Seed is produced at higher altitudes in the hills.
- Seed to seed method
- The first season bulb crop is left to over winter in the field, so as to produce seed in following season.
- Bulb to seed method
- Bulbs stored in previous season are lifted, selected and replanted to produceseed in 2nd year.
- Mostly bulb to seed method is followed because
- It permits selections of true to type and healthy bulbs for seed production.
- Seed yields are comparatively high.
- Sowing of over wintered bulbs is done from Middle of October to middle of November
- Apply 200q/ha of FYM, 60 kg each of N and P and 80 kg of K per hectare.

Chapter 6

Root Vegetables

Carrot: *Daucus carota* L; Umbelliferae

- ☆ Chromosome number - 2n= 18
- ☆ Common name- Poor mans ginseng
- ☆ China is the leading producer of carrot
- ☆ Cool season vegetable crop
- ☆ Carrot root rich in sucrose 10 times more than glucose and fructose
- ☆ Kanji appetizing drink is prepared from Asiatic carrot
- ☆ 1st male sterility reported in carrot (Welche and Grimble)
- ☆ 1st reported heterosis breeding Poole 1937
- ☆ Carrot seed is commercially used in France
- ☆ Carotenoid range from 80-120ppm
- ☆ Beta carotene per cent in carrot root 50 per cent
- ☆ Anthocynin content of black carrot range 1750mg/100gm
- ☆ Hispid branches produce from second year
- ☆ Queen Anne's lace wild carrot
- ☆ Flower- Apiquinous
- ☆ Objective of breeding-
- ☆ Bolting of bigger problem
- ☆ Non bolting type- mass cell, Pedigree method

- ☆ Pusa Kesar
- ☆ Kuroda
- ☆ Kashmir Beauty, Golden Heart var. of carrot
- ☆ Flavour of carrot by glutanic acid
- ☆ Short duration crop
- ☆ Cultivated spp- *Dacus carrota* (*Daucus sativus*)
- ☆ Edible portion of carrot enlarge- Fleshy tap root
- ☆ Conical type of roots.
- ☆ Rate of respiration- Moderate
- ☆ Better perform in pH 5.5-6.8
- ☆ Crude fiber 1.2g/100g
- ☆ 71.7 per cent cellulose, hemicellulose 13.0 per cent, lignin 15.2 per cent
- ☆ Carotenoid range 6000-54800μg/100g
- ☆ Carrot gives 48 callories/100g
- ☆ Level of inbreeding depresion- very high
- ☆ One gram of carrot seed contains 850 seeds
- ☆ Male sterility is common in carrot
- ☆ Leaf hopper and cut worm are serious insect pest of carrot
- ☆ One single plant produces 50000 seeds
- ☆ Iso-coumerin is responsible for bitter taste
- ☆ Root to seed method of planting material stecking
- ☆ Anthesis of single umble complete in 7-9 days
- ☆ Crown greening of carrot is due to exposure of root to
- ☆ Spartan Bonus- Three way cross hybrid
- ☆ Fruit type – schizocarp and seed (Compound umble)
- ☆ Phyto nutrient found in carrot
- ☆ Asiatic type- High yield but low carotene contains
- ☆ European type- Low yield but high carotene contains
- ☆ Cavity spot disorder is due to Ca and excessive uptake to potassium
- ☆ Leaf spot late blight, bacterial wilt, aster yellow, soft rot are common disease of carrot
- ☆ Carrot entomophillus in nature
- ☆ European carrot orange colour, Asiatic colour red and purple colour

- ☆ Tropical/asiatica/oriental/annual type: Pusa Kesar, Pusa Meghali
- ☆ Cortex colour: Red- Pusa Kesar, Local red, Pusa Rudhira
- ☆ Orange red- Nanteney
- ☆ Ornage- Pusa Yamdagini, Ooty-1, Chateney, Imperator, Pusa Nayanjyoti
- ☆ Black variety- Pusa Ashita (IPC126)
- ☆ Punjab black beauty- (PCB-5)
- ☆ Ovule var.- Early Scarlet Horn
- ☆ Brown anther type-BA, is present in all cultivates orange
- ☆ Colour open pollination var.
- ☆ Brown anther type 1st found in (Tender shoot var.)
- ☆ Petaloid stamen type - 1st found in (Cornell wild carrot)
- ☆ It is commonly used in hybrid seed production
- ☆ Pusa Kesar - Local red × Nantes half long- rich carotene -38mg/100gm
- ☆ It is seed production in plain area
- ☆ Pusa Meghali- Pusa Kesar × Nanthes- highest carotene- 1157IU/100mg
- ☆ Nanthes- European seed production in hilly area
- ☆ Kuroda- Sel 233(nanthes × Sel.29)
- ☆ Sel29, Sel-233, PAU Ludhiana
- ☆ Ratio of female and male for hybrid seed production4: 2
- ☆ Temperate or europian type
- ☆ Nanthes half long, coreless, early nanthes
- ☆ Chantenay- excellent cultivar for canning and storage
- ☆ Blunt type- Danvers- mid season cultivar suitable for fresh market and processing
- ☆ Zeno- Introduced from Germany
- ☆ Ooty 1- single cross hybrid of carrot (Spartan North)
- ☆ Pusa Yamdagini- EC9981 × Nantes (Sel no. 5)
- ☆ Three way carrot hybrid (*spartan classic*) (*spartan premium*) (*Spartan winter*)
- ☆ Optimum temp for root formation 18-22 ^{0}C
- ☆ Germination temp 7.2^{0}C-23.9^{0}C
- ☆ Seed rate- 5-6 kg/ha
- ☆ Carrot root fly incidence can be reduce- By intercropping- shallot-or-chive

- ☆ Forking- Hard soil does not allow (Straight growth)
- ☆ Maximum concentration of carotene 90-130DAP
- ☆ Splitting – sudden change in soil moisture
- ☆ Yield- 20-25tonnes/ha
- ☆ Optimum temperature for storage 0-4.4 ^{0}C, RH 93-98 per cent
- ☆ Fruit- 3mm long
- ☆ Red rooted cultivar- Kintoki
- ☆ Optimum temp. for colour development 15-21 ^{0}C
- ☆ Family breeding is type of recurrent selection
- ☆ Best temp. for orange root colour developed 15.6-21.1 ^{0}C
- ☆ Root shape governed by 3 gene DND
- ☆ Petaloid male sterility Munger -1953
- ☆ Pro-vitamin found in carrot
- ☆ Carrot is gross feeder of potash
- ☆ Lygus bug is serious pest of carrot
- ☆ Pusa Nayanjyoti-F_1 hybrid- 1st temperate carrot hybrid using CMS
- ☆ Round var.- French forking
- ☆ Pusa Vashudha- Red type 1st tropical hybrid using-CMS
- ☆ Hisar Garlic by mass selection
- ☆ Pusa Barsati- Tolerant to heat and humidity and suitable for kharif sowing
- ☆ Prima- by recurrent selection
- ☆ Pusa Rudhira- self colour core red variety tropicaans Asiatic
- ☆ B-11: Texas gold spick
- ☆ Pusa Pyasha- Murford colour variety
- ☆ Carrot seeded contain germination inhibitore carrotal
- ☆ Impretore- Nantes × Chenthes- mid season to late maturing
- ☆ Oxheart-hard shape carrot cultivar, beta carotene in carrot 50 per cent
- ☆ Seed yield-500-600 kg/ha
- ☆ Carrot is high cross pollinated due to andromonoecious
- ☆ Carrot seed will not germinate 3.9 ^{0}C
- ☆ Optimum spacing 25-30 cm×8-10cm
- ☆ Optimum plant population of carrot 100-130plant/m^2
- ☆ Per ha yield of temperate type carrot 100-150q/ha

- ☆ Carrot yellow disease by virus
- ☆ The carrot crop needs deep loose loamy soil.
- ☆ It requires a soil pH range of 6.0 to 7.0.
- ☆ Season; Hills: At elevation above 1500 metres, carrot can be grown throughout the year under assured irrigation.
- ☆ At elevations between 1000 – 1500 metres. Carrot can be grown in July – February; Plains: August.
- ☆ Seed rate: 4 kg/ha. Mark the rows with a spacing of 25 – 30 cm., Sow the seeds mixed with sand (one part of seed with 4 parts of sand).
- ☆ Thinning: Hills: 10 cm between plants; Plains: 5 cm between plants.
- ☆ Preparation of field; Hills: Prepare the land to a fine tilth and form raised beds of one metre breadth and convenient length
- ☆ Plains: Two ploughings are given and ridges and furrows are formed at 30 cm spacing.
- ☆ Irrigation: Once in five days.
- ☆ Application of fertilizers: 30 t of FYM and 90: 90: 90 kg of NPK per ha as basal dose and 45: 45: 45 kg of NPK after 45 days. Apply 25 kg of $ZnSO4$/ha as basal.
- ☆ After cultivation: Spray Fluchloralin 1 l a.i./ha immediately after sowing the seeds to control weeds or first weeding to be done on 15th day.
- ☆ Thinning and earthing up should be given on 30th day.
- ☆ Splitting of roots is a physiological disorder which is usually seen when there is a sudden increase in soil moisture after prolonged drought.
- ☆ Forking is another phenomenon where in the hard soil does not allow the straight growth of tap root which results in formation of forked root.
- ☆ Forked roots and split roots fetch poor price in market.
- ☆ Keeping the soil moisture at optimum condition will help to keep the soil also loose thereby help to avoid split and forked roots.
- ☆ Plant protection: Carrot is not much affected by pests.
- ☆ Nematode: Application of neem cake @ 1 ton/ha at planting to control root knot nematode, *Meloidogyne spp.*
- ☆ Diseases; Leaf spot: Spray Mancozeb @ 2 g/l.

Radish: *Raphanus sativus* L; F.- Brassicaceae

- ☆ Chromosome no- 2n=2x=18
- ☆ Origin- Mediterranean region
- ☆ Pod type –Siliqua
- ☆ Type of inflorescence –Terminal raceme
- ☆ Modified form of fleshy root of radish is a **fusiform**
- ☆ Common name –Muli, Mulamulagi
- ☆ Manual harvesting is common in radish
- ☆ Root cum leafy vegetable suitable and sub tropical climate
- ☆ Preparation of radish curing, lever for gall bladder problem
- ☆ Roots are used in treating in piles
- ☆ Juice of fresh leaf duratic and laxative
- ☆ Pungency in radish is due to presence of **glucosinolate (Isothiocyanate)**
- ☆ **Glucose** is the major sugar present in the radish roots
- ☆ Edible portion of radish root developed from both primary root and hypcotype
- ☆ It is cross pollinated vegetable due to spherophytic system of incomatibilty
- ☆ Mode of pollination entophillus
- ☆ Objective of crop improvement should be not wilting
- ☆ **Takoan** is a famous Japanese radish pickle
- ☆ **Asiatic var.-**
- ☆ Pusa Desh- Selection from local material
- ☆ Pusa Rashmi- Suitable for main season- tolerant to high temperature
- ☆ Pusa Chetki-It is selection from Denmark- suitable for growing in hot month
- ☆ Japanes White- IARI- introduced from Japan
- ☆ Punjab Safed- A derivation of cross White 5×Japanes White
- ☆ CO-1: TNAU root are milky white suitable for growing in plain in all season it is also set seed in plain
- ☆ Radish seed diameter 3mm
- ☆ Arka Nishant- Improvement over collection IIHR-72 from Singapur developed multi resistant at IIHR

- ☆ Resistant pithiness, premature, bolting, root branching and forking it set seed freely in plain
- ☆ Chinese Pink- Introduced identified by Dr Y.S.P.R Nauni- Dual season var.
- ☆ Temperate reason var.-
- ☆ White Icicle- A table var., 30 days for harvest
- ☆ Pusa Himani- Black × Japanes White
- ☆ Hilly var. throughout of year Nov-Jan
- ☆ Asiatic var.- Originated from *Raphanus sativus raphanistroid*
- ☆ Botenic var. of radish was given by Banga 1976
- ☆ Rapid Red White Tiped- Table var.- 21 day of harvest
- ☆ Scalet Long- root long tapering to a point with- red skin
- ☆ Root developed best flavour, texture,size 10-15°C
- ☆ Root also becomes hard pithy and pungent during hot weather
- ☆ Pink, colour is due to presence of **anthocyanin**
- ☆ **Purple colour- cyaniding**
- ☆ **Red colour due to presence of -Pelargonidin**
- ☆ North hill sowing is done from 1st forth night of march and till Late Octoberor beginning of November
- ☆ Asiatic type (August - January)
- ☆ Europianty var.-Sept.- March
- ☆ South Indian hill -April- June
- ☆ Plain- October- December
- ☆ Seed rate 9-12 kg/ha
- ☆ Tropical 8-10kg/ha
- ☆ Temperate 10-12kg/ha
- ☆ Seed hardening is done at 22°C Asiatic type
- ☆ Asiatic type 45cm on ridged, europian type 5-10×3cm
- ☆ **White rust resistant var.- Arka Nishant**
- ☆ Hallow heart formation high temperature 16-30 days of sowing
- ☆ **Akashin-**Boron deficiency, it is high day and night temperature
- ☆ Soil -20 °C as well as by low soil moisture
- ☆ Bitterness of root is more in summer crop than spring or autumn
- ☆ Radish root become ready to harvest 25-55 day

- ☆ Root yield- 15-20 tonnes/ha – Indian var.
- ☆ 5-7 tonnes/ha Europian var.
- ☆ Storage temperature 0-1 ^{0}C, RH 90-95 per cent for 2 months
- ☆ Garden/europian/occidental/small coll season- ***R. radicula***
- ☆ Oil seed/fodder radish- ***R. var. Oleiformis***
- ☆ Rat-Tail/Mougri Radish- ***R. sativus var. candatus***- Pod are use as vegetable in India
- ☆ Large radish/black/spainsh radish- ***R. sativus var. nigra***
- ☆ ***Indian- R. indica***
- ☆ Japanese Chines/**winter- *R.sativus var. raphanistroids***
- ☆ Chinese radis-***R. sativus var. longipinnatus***
- ☆ Horse radish- ***Amoracia rusticana*** 2n=32
- ☆ Good source of vitamin C
- ☆ Fleshy root radish- Fusiform root
- ☆ Famous Japanese radish pickle- Takaon
- ☆ Radish pungency- isothiocynate
- ☆ Red Purple Pink- Anthocynin
- ☆ Purple cynanidin
- ☆ Red-pelargonidin (Raphamism)
- ☆ Pink skin radish Vit. C
- ☆ Inflorescence terminal raceme
- ☆ Male sterility in radish 1st reported in Japanes radish by (Ogura) 1968
- ☆ Radish CMS system- Ogura type- hybrid seed production
- ☆ ORF-138 (ogura CMS), (Mitochondrial gene responsible)
- ☆ Radish polyploidy is seed borne disease- Kashi Sweta, Kashi Hans
- ☆ New var. Palam Hriday, Pusa Yamini, Pusa Gulabi (recd var.), Pusa Mridula (extra early and table purpose)
- ☆ Green sholder radish var.- Pusa Sanka
- ☆ 1st purple fleshy var.- Pusa Sagarika
- ☆ Long stem var.- white colour, Pusha Bidhu
- ☆ Popular north Indian (hills queens)
- ☆ Soil: Sandy loam soils with high organic matter.
- ☆ The optimum soil pH is 5.5 to 6.8. Roots obest size, flavour and texture are developed at about 15 ^{0}C.

- ☆ Season of sowing: June –July in hills and from September in plains.
- ☆ Preparation of field: The land should be prepared to fine tilth and levelled.
- ☆ Spacing: 15 cm x 10 cm.
- ☆ Irrigation; Plains: Once in a week.
- ☆ Application of fertilizers: Plains: Apply FYM at 25 t/ha and 25 kg N, 100 kg P and 50 kg K/ha as basal dressing and 25 kg N/ha after 30 days.
- ☆ Fertigation: Fertigation requirement: 50: 100: 50kg of NPK/ha. Apply once in three days Spacing: 30x15x10 cm in paired row system.
- ☆ A total of 10 rows is accommodated in one bed.
- ☆ After cultivation: Weeding and hoeing should be done at the second weeding, thinning of thickly sown plants should be done.
- ☆ Pests Aphids, flea beetles and mustard saw fly: Spray Malathion 50 EC 1 ml/l twice or thrice at 10 days intervals.
- ☆ Diseases; White rust: Spray Mancozeb 2 g/l or Copper oxychloride 2 g/l.

Beet Root: *Beta vulgaris* L.; Family-Chenopodiaceae

- ☆ Varieties: Ooty 1, Crimson Globe, Detroit Dark Red and Red Ball.
- ☆ Soil: It comes up well in all types of friable soils.
- ☆ The suitable soil pH is 6.0-7.0. It is considered to be a cool weather crop.
- ☆ Season and sowing: July – August.
- ☆ Seed rate: Sow 6 kg/ha directly in ridges at a spacing of 10 cm.
- ☆ Preparation of field: Plough to a fine tilth and form ridges and furrows 30 cm apart.
- ☆ Irrigation: Irrigate the field copiously immediately after sowing and afterwards as and when necessary.
- ☆ Spacing: 30 x30 x10 cm as four rows in each paired row/raised bed system.
- ☆ After cultivation: 20 days after sowing thin to a single seedling per hill.
- ☆ Plant protection; Pests: - Leaf miner and flea beetle: Spray Malathion 50 EC 2 ml/l.
- ☆ Diseases; Cercospora leaf spot: Spray Mancozeb at 2 g/l. Rhizoctonia root rot: Spot drenching with Carbendazim @ 1 g/l.
- ☆ Yield: 20 - 25 t/ha in 120 days.

Turnip; *Brassica rapa* Linn Syn. *Brassica campestris* var. *rapa*; Family- Cruciferae

- ☆ Origin: Indo-China
- ☆ Chromosome No=2n=2x=20
- ☆ Grown mainly for its enlarged roots
- ☆ Young leaves are rich source of ascorbic acid, iron and vit- A
- ☆ Flashy thicken underground portion of turnip is actually the hypocotyle
- ☆ Colour bellow ground portion white and yellow
- ☆ Above ground portion red purple white yellow
- ☆ Roots attain edible maturity in (40-80 days)
- ☆ Inflorescence (Terminal raceme) on the main stem
- ☆ **Mass selection is major breeding programme**
- ☆ Turnip has strong sporophytic self-incompatibility
- ☆ It no need for emasculation
- ☆ Thinning is important operation
- ☆ NaCl and CO_2 can used to overcome SI
- ☆ Turnip classification given by Shoemaker (1949)
- ☆ Optimum soil pH 5.5 to 6.8
- ☆ It can tolerate frost and freezing temp.
- ☆ Temperature bellow 10 ^{0}C induced flowering
- ☆ Most favorable weather for development of root (10-13 ^{0}C)
- ☆ Air temp- 18-23 ^{0}C
- ☆ In north India the sowing is taken up from September-December)
- ☆ Lowe hill July to October; higher hill July to September
- ☆ Seed rate- 3-4 kg depth 1.5 to 2.5 cm
- ☆ Turnip is napiform roots
- ☆ Multigerm found in turnip
- ☆ Turnip is closely related to Sweede rutabaga sweet turnip
- ☆ ***Brassica napus* var. *napo brassica*** = Amphidiploids 2n= 38, AACC genome ***B. compestris x B. oleraceae***
- ☆ Honey bee are chief pollinator agents (Rapa)
- ☆ Diameter during harvest of turnip 5-7.5 cm root

- ☆ Dry matter contain should (8-9 per cent)
- ☆ Viabilitiy of turnip seed (5-6 year), Test weight- (3.3 gm)
- ☆ Yield 25 t/ha and seed yield 550kg/ha.
- ☆ Major problem in seed production phyllody disease
- ☆ Club rot resistant turnip: Manga from New Zealand
- ☆ Asiatic/Tropical type: Pusa Kanchan (Local Red x Golden Ball) red colour, seed set in plain
- ☆ Pusa Swati –Pure white
- ☆ Punjab Safed- pure white sowing in Oct.
- ☆ Temperate type: Pusa Swarnima- Japanese White x Golden Ball
- ☆ Pusa Chandrima- Japanese Whitex Snowball hybridization b/w (Asiatic and European)
- ☆ Pusa White, Golden Ball (extremely susceptible to phyllody), Snowball (55 days)
- ☆ Early Millon Top Red - Purplish Red at top and white in lower portion, extra early high yield cultivar. 45 days
- ☆ Purple top white globe high yield large rooted cold season

Chapter 7

Leguminous Vegetables

Peas: *Pisum sativum* L.; Fabaceae

- Chromosome number- 2n=14
- Origin- Central Asia
- Inflorescence- Raceme
- Type of fruit- Legume pod
- Isolation distance- Foundation 20m, and certified-10m
- Pea is a herbaceous winter annuals
- Ancestorer of pea-*Pisum elatius*
- Protein- 25 per cent
- Proteins contain in pea- 7gm/100gm
- Field pea- ***Pisum sativum var. arvense*** (frost registant)
- Edible podedvariety- ***Pisum macrocarpam***- Snap pea
- Dwarf pea- ***Pisum humile***
- Phenotypic performance is good for seed selection
- Red yellow pea-***P. fulvum***
- Pea seed also called- ovule
- Maturity- Early type 50 days
- Mid type- 60-65 days
- Late type- 70-75 days

- Most important growing state of garden pea- UP 60 per cent
- Production, productivity- AP
- World famous garden pea breeder- S. Blixt (Swedan)
- Antinutrient factor- Cynide
- Pod maturity measured by- Tendrometer (Heat unit system)
- Optimum temperature for growth- 13-19 ^{0}C
- Optimum temperature for pea seed germination 22 ^{0}C
- Seed can germinate minimum temperature 5 ^{0}C
- High dry weather interferces- pollination/seed setting
- Effect no. of pod per plants/lower quality of pod produce due to conservation of (sugar to hemi-cellulose + starch)
- Seed rate- Early 100-10 kg/ha, Late 80-90kg/ha
- 10ml of 10m- CCC- The five nodes stage growth and yield
- Spacing- 30cm row×row; 5-10 cm plant×plant
- Seed are sown about -2.5cm deep
- Weed control
- Pre sowing- 3lit.ter/ha – Basaline
- Atrazine/propazine/simazine- 1.5kg/ha
- Promatryn- 1kg/ha
- **Mutation breeding worked in pea**
- Pre plant bear- Tendril
- Natural habit of plant- climb
- Growth regulator-15 ml of 10ml solution CCC growth and yield
- Improvement of garden pea- started in 1940
- Harvesting is delayed the pod surface become- coarse
- Qualitiy also reduces fast if the harvested pod areleft in the field.
- Acacia mutant- tendril converted into leaflet responsible gene (af)
- Yield
- Early var.- 2.5-4.0 tonnes/ha
- Mid var.- 6.0-7.5 tonnes/ha
- Late var.- 8.0-10.0 tonnes/ha
- **Shelling per cent range 35-50 per cent**
- Critical stage-grain feeling period

- ☆ Gypsum and pyrite- Complete the deficiency of sulphur
- ☆ Storage temperature 0°C and RH 95 per cent
- ☆ Filthum first and meteor resist. sources of pea
- ☆ Fresh should be removed before the peas are packed
- ☆ If sown early- suffer from wilt
- ☆ Late sowing attack P.M and D.M
- ☆ Wrinkled seed suitable for- freezing and canning, Smooth for- dehydration
- ☆ Protein contains- Smooth seed- 23-31 per cent (Highly starch)
- ☆ Protein contains- Wrinkled seed- 33 per cent
- ☆ Early variety packaging in 2 times and mid or late-3-4 times
- ☆ Multi resist. line JP-179, JP-9, JP Botri brown 3 and 4
- ☆ Snap pea- ***P.S. macrocarpam***, thick pod well
- ☆ Snow pea- china pea- ***P.S. saccharanthum***, thin pod well
- ☆ Round seeded var.- Highly starch contain
- ☆ Seed yield 1500-2000kg/ha
- ☆ Viabilitiy of garden pea- 2-3 year
- ☆ Sowing time in north plain- Oct- mid Nov.
- ☆ Prolonged application of DAP cause sulphur deficiency
- ☆ Insect pest-
- ☆ Pea pod borer- ***Helianthus sp.***
- ☆ Control- 0.2 per cent thiourea at flowering
- ☆ Pea life minor- ***Agromyza flaveola***
- ☆ Control 0.1 per cent monocrotophus
- ☆ Powdery mildew- ***Eryshiphe poligoni*** 0.1 per cent dinocap
- ☆ Root rot- ***Rhizoctonia* spp.**
- ☆ Control- 0.2 per cent cat of bavestin
- ☆ **Morsh spot**-Due to Mn deficiency/micro nutrient disorder
- ☆ Mosaic disease transmitted by- Aphid
- ☆ Best method of processing- freezing/air blast freezing
- ☆ Popular method of canning
- ☆ Jawahar matar-4
- ☆ Variety- Early var.- Alaksha- 1st early canning var.
- ☆ Pusa Pragati- resistant to P.M

- Arkel- Introduced from England suitable for dehydration
- Jawahar Matar- 3 and 4; Pant matar -2.
- Hissar Harit- Bboniville X P-23
- Ageta-6, Asauiji, Kasha Nandini
- Harbhajan- Extra early var. 45 days
- J.M.-54 dual cross var.
- Mid season- Bonneville- Introduced from U.S.A.
- Lincon- Introduced from- U.S.A.
- Jawahar matar, Pant uphar, Punjab-88 Mithiphali, J.P.-19- Resistant to moderate P.M
- VL-3
- Rust- Lincon and Windo- branching habbit
- Canning var.- Early badger/Alaksha (Bluish green colour seed) resistant to fusarium/heat drought
- Resistant to P.M. and rust- Arka Kartik/Arka Sampoorna/Palam Priya
- Riaistant to P.M-Arka Ajit
- Resistant to wilt- Delwinchecamondo/New Era
- Fusarium wilt-Alaksha/Surprise
- Wrinkle seed var.- Arkel/Bonneville/Early Badger
- Smooth seed var.- Asaujii/Meteor
- Most popular variety in M.P- Khaprkheda double podedvar.
- JP-19/Salvia- Edible poded var. Mater Ageta-6
- Early December- Bushy var.
- NP-29- IARI var.
- Little Marvel- Introduced- England- Chelsea×Alaksha
- Lucknow Bonia- 1st variety come to Delhi market
- Recent Rachna is P.M. resistant var.- T163×T10
- Thomas laxton- Most suitsble for freezing
- Early jaint/Aldaman/linco- Hill var.
- Alderman- Exilent for home garden double podded var.
- Solan Nirog- mid var. and resistant to P.M.
- Ooty-1 resistant to white fly
- Smooth seeded main season- Kannari

- ☆ JP 83- (Arkel×J.P829)×(46C×J.P 501)- Resistant to P.M
- ☆ Perfection new line var.
- ☆ J.P-71 (Arkel×SP827)×(JP 501×JM-1)- Resistant to P.M
- ☆ Mutant var.-Saria
- ☆ Duke of albony- Also knowns as telephone need stacking
- ☆ Produce 1st flower 12-14 nodes
- ☆ Mature seed of pea and bean contain- Phytic acid
- ☆ Toxic substance- Trypsin inhibitor
- ☆ Acid present- Citric acid
- ☆ Enzyme- Hexa phosphoric acid
- ☆ P.M in pea controlled by single recessive gene
- ☆ Short day plant
- ☆ Respiration rate is high
- ☆ Soil salinity- Less tolerant
- ☆ Lime requirement- Moderate
- ☆ Pineor in producing garden pea variety- Andrew knight
- ☆ Starch present in the form of amylopectine in round seed
- ☆ Amylases present in wrinkled seed
- ☆ Pea and bean are self pollinated crop due to kell petal
- ☆ If peas are grown- earlier- then suffer from wilt, if letter mildew disease
- ☆ Pea area 5646000 ha, 5694000MT production
- ☆ Green arrow resistant to fusarium wilt
- ☆ Freezing temperature pod- (10 ^{0}C)
- ☆ Suitable for freezing -Thomas laxton, Alderman, Oregone-523
- ☆ Difference between freezing and chilling
- ☆ Canning purpose var.- Early badgen, Lincon, Alaksha-81
- ☆ Chilling temp 0-1°C
- ☆ Suitable for dehydration- Arkel, NP-29, Apex
- ☆ Freezing temp- (-1 ^{0}C)
- ☆ Breeding goal- High green pod
- ☆ 9-12 pod will be required
- ☆ Important doner of – Earlyness Asuji, Lucknow Bonia, EC-3
- ☆ More number of pod- PL-26, 50,56

- Long pod EC-179
- Molecular marker gene- ADH-1 marker
- RAPD marker- Y-14, Y-15
- Edible poded var.- ***Pisum sativum macrocarpoum***
- Sugar Born, Honey Pod, Argon Sugar Poded, Mithi Phali
- High temperature to towards maturity- yield loss and pod due to conversion of sugar to hemicellulose
- Salt treated pea retain chlorophyl better during storage
- Pea is very perishable
- Shulter breeding used in pea
- Early cultivar-Heat unit 1534
- Late cultivar-Heat unit 3942
- **Shelling per cent of pea**
- Alaksha- 42 per cent
- Arkel-40 per cent
- Early Badger-36 per cent
- Little Marvel-40 per cent
- NP 29-50 per cent
- Soil: Well drained loamy soil with optimum pH range of 6-7.5.
- Season and sowing: Sow the seeds during February- March and October - November in plains.
- Treat the seeds with ***Trichoderma*** @ 4 g/kg or Thiram or Captan @ 2 g/ kg of seeds to avoid seed borne diseases. Treat the seeds with ***Rhizobium*** culture @ of 2 kg and apply 2 kg Phosphobacteria as soil application just before sowing.
- Preparation of field: Dig the land thoroughly fine tilth.
- Spacing: 45 cm x 10 cm.
- Seed rate: 100 kg/ha.
- Irrigation: Once in a week.
- Application of fertilizers: Apply FYM @ 20 t/ha and 60 kg N, 80 kg P and 70 kg K/ha as basal and 60 kg N/ha 30 days after sowing.
- After cultivation: Weeding should be done 15 days after sowing.
- Subsequent weeding as and when necessary. Stake the plants on 30th day of sowing.
- Harvest: Harvesting can be done on 75 days after sowing.

- High temperature during harvest affects the quality of peas.
- Plant protection
- Pests; Pod borer: Spray Carbaryl 50 WP thrice at fortnightly intervals @ 2 g/lit. Dust with Carbaryl 10 D at the rate of 25 kg/ha.
- Aphids: Spray Methyl demeton 25 EC or Dimethoate 30 EC or Monocrotophos 36 WSC or Phosphamidon 40 SL @ 1 ml/l of water.
- Diseases; Powdery mildew: Spray wettable sulphur @ 2 g/l or Dinocap @ 1 ml/l or Tridemorph @ 0.5 ml/l or dust sulphur @ 25 kg/ha three rounds at 15 days interval.
- The edible-pods should be uniformly bright green (light to deep green but not yellowgreen), fully turgid, clean, and free from damage.

Cowpea: *Vigna unguiculata* (L) Walp; Fabaceae

- Chromosome number-2n=22
- Common name- China pea/Black eyed pea/Kaithir pea/Southern pea
- Immatured pod used as vegetable
- Day natural plant
- It is a hardy crop
- Shallow rooted
- Rainy season veg.
- Self pollinated crop
- Inflorescence- Multiple raceme
- Proginator- ***V. anguiculata var. mensensisii***
- Flower colour white, 2-3 flower on each peduncle
- Protein contains 22-23 per cent
- Common method of breeding- pedigree method
- Rainy season 12-15kg/ha
- Summer seed rate 25kg/ha
- 1st time inoculation of seed with rhizobium sp.
- Weed control- flochloraline 2lit./ha
- Harvesting 45 day after sowing
- Cow pea also called vegetable meat
- It is also known as catch crop, drought tolerant and sensitive tolerant, water logging
- Botanical name of Catjang bean- ***Vigna catjang*** 7-13cm long, pod(kidney shaped)
- Asparagus bean/yard long bean-***Vigna sesquipedal***
- Southern pea- ***V. anguiculata***
- Cow pea has strong tap root system
- Spraying of MH 50-200 ppm before flowering improve pod set and yield
- Yield 50-80 q/ha green pod
- 12-15q/ha seed yield
- Proteage inhibitors have been reported and (oligosacharied).
- Insect- pod borer- endosalfon/carbaryl- 2gm/1lit.

- ☆ Aphide- phosphomidon- 0.5ml/lit.
- ☆ Thrips can cause yield loss upto- 100 per cent
- ☆ Powdery mildew/rust- sulpher 3gm/litre, dinocap 1ml/lit.
- ☆ Photo insensitive crop- both season summer and rainy season
- ☆ Anthracnose/Septoria blight- dithan M-45, 2gm/lit.
- ☆ Pods are rich in protein/Vit-B/minerals
- ☆ Vit-A contain 941 IU/100gm
- ☆ Biflora pods- use smaller 7-13 per cent and kidney shaped seed
- ☆ China Red- Resist. to phytopthora stem rot
- ☆ Var.-
- ☆ Pusa Rituraj- Double purpose, photo insensitive
- ☆ Pusa Komal- Highly rasist. to golden yellow mosaic virus
- ☆ Pusa Phalguni- Bushy plant/dwarf var./Introduced from philipins
- ☆ Spring season var. Feb-March
- ☆ Biflora- Resistant to pod borer
- ☆ Pusa Barsati- Introduced from Philippians/rainy season
- ☆ Pusa Dofsli-Bushy plant/both season var. summer/spring
- ☆ Pusa Komal- Bushy/both season/bacterial resist var.- pureline
- ☆ Arka Garima- Bushy plant/photo insensitive- backcross
- ☆ Arka Suman- Bushy plant/photo insensitive, resistant to Rust
- ☆ Arka Samridhi- Bushy plant/photoinsestive
- ☆ Kashi Shymal, Kashi Sudha, Kashi Kanchan, Kashi Unnati, Kashigauri
- ☆ Narendra Lobia -1 Bushy plant/bold seed
- ☆ Cowpea-263 bushy/Dwarf/medium green pod
- ☆ C-153 resistant to root knot nematodes
- ☆ Bhagya Lakshmi- Anthracnose resistant
- ☆ Yard long bean- late maturity var.
- ☆ Pusa Komal- Pusarituraj×P246
- ☆ Bidhan Barbati1- resist to cowpea mosaic (cowpea golden mosaic)
- ☆ Bidhan Barbati-2 semi determinant
- ☆ Grain type- Double purpose
- ☆ Grow wide time 21-35 ^{0}C
- ☆ Germination is better at 12-15 ^{0}C

- Better growth development 21-27 ^{0}C
- Sowing time- rainy season June- July 45 cm×60 cm R×R, 15cm P×P
- Summer season: Feb-March 30 cm×30cm R×R, 15cm P×P
- Pusa Rituraj- Pusa Dofasali×EC26412
- Asseem- Pusa Phlguni× EC21622
- Progenitor**- *V.A mensensii*** (sub species)
- Drought hardy crop 21- 31 june July
- Seed rate- rainy 10-12kg/ha
- 22-25kg/ha
- 25 kg/ha
- Spacing 30×45cm – 45×60 RR
- PGR Mallik hydrazide (MH) - Pod yield
- NAA- @15ppm Sinduce fruit set
- Harvesting for green pod- 45 day after sowing
- Yield- green pod 40-80q/ha
- Test weight 22 gm
- Isolation distance
- 50 m- foundation
- 25 m- certified
- Storage temperature 0C 85-90 per cent RH for 15-20 days
- Soil: Well drained soil with high organic matter.
- Season: June – July (Rainfed), February – March (irrigated).
- Treat the seeds with 600g of ***Rhizobium*** bacterial culture before sowing as in cluster beans.
- Dibble the seeds on both sides of the ridges or in lines in the beds.
- Preparation of field: Plough the field to fine tilth. Form ridges and furrows at 45 cm apart or beds of convenient size.
- Irrigation: Give irrigation immediately after sowing and on 3rd day, thereafter once a week.
- Application of fertilizers: Apply FYM 25 t**, *Azospirillum*** @ 2 kg and Phosphobacteria @ 2 kg/ha, and N 25 kg and P 50 kg/ha for irrigated crop.
- Apply FYM at 12.5 t/ha and N 12.5 kg and P 25 kg/ha for rainfed crop.

- Fertilizers can be applied in several split doses at fortnightly intervals.
- Pinching: Before flowering, the tendrils should be pinched thrice for getting bushy plants.
- After cultivation: Give one hoeing and weeding on 25th day after sowing.
- Pests; Aphids: Spray Dimethoate 30 EC 1 ml/l or Methyl demeton 25 EC 1 ml/l.
- Diseases; Powdery mildew: Dust sulphur 25 kg/ha or spray Wettable sulphur 2 g/l.

Lab Lab/Dolichos Bean: *Lab labpurpureus* var. *typicus*; Fabaceae

- Chromosome number 2n=22/24
- Most ancient among cultivation plant
- Common name- Dolichos bean/Hyacinth bean/Lab-lab
- It is herbaceous perennial
- It is two type-1- typicus garden type-***Dolichus purpureus***
- 2- Dolichos lignose- High fibre content
- Short day plant
- Seed rate- 20-30 kg/ha
- Sandy loam and clay loam soil are best
- It is photosensitive crop
- (CO) variety found in Indian bean
- Flowring and pod development period are the critical stage dependent on the atmospheric condition
- Deutorious effect- Minimised by removing seed coat
- Weed control- Fluchlorine at the rate 2 litre/ha.
- Average yield 50-80q/ha
- Temperature -18-20 ^{0}C
- Bunch type ready two month
- Pole type three month for 1st harvest
- Toxic substance- Trypsin inhibitor, it is present in mature and coloured seed (Anti-nutritional factore)
- Var.- Arka jay- Photo insensitive and resistant to moisture stress/heat
- Annual railfall- 630-890 mm
- Arka Vijay- Dwarf, and Photo-insensitive/double purpose
- Pusa early prolific- Pole type/early var./suitable for autum and spring
- Pusa sem. 2, 120-125 days maturity- Highly torerant to anthracnose/virus/pod borer/photo sensitive
- Pusa Sem-3 is Photo sensitive
- Blue Lake- early var./Bushy/Photo insensitive/White flower/Dual purpose
- Konkan Bhusan- Photo insensitive/Dual purpose

- ☆ Dasarawal- Deepaliwal
- ☆ Arka Jay/Arka Vijay- It can be grown through out the year
- ☆ Climbing var.- Rajni/Deepali/kdb-403,405,71
- ☆ Pusa early Rajni- Improved var.
- ☆ CO10- mutant var.(CO-9-CO-12)
- ☆ Wal Konkan-resistant to yellow mosaic virus, dual purpose
- ☆ Vijay, vahi, habula var.-1,2,3 dual type
- ☆ Sowing time- June July- north India
- ☆ July-August- south India
- ☆ Arka Amogh/Arka Komal/Kasha Haritima/Arka Sambhram
- ☆ Arka Saumya/Arka Bold= All var. are pole type
- ☆ Best temperature for seed germination- 18-27 ^{0}C
- ☆ Bush type- better growth and development 21-27 ^{0}C
- ☆ Better fruit set 12-18 ^{0}C
- ☆ Arka Jay- suitable for vegetable purpose
- ☆ Dolichos derived from- Greek word
- ☆ Lab-lab derived from Arabic word
- ☆ Hebbal var. Bush type- Pods are not edible but used hybridization
- ☆ Light irrigation- Very successful and luxurious growth
- ☆ Major problem- Flower drop 10-12 per cent flower develope
- ☆ Control $CaCl_2$ 0.1 per cent, NAA 100ppm
- ☆ Moisture stress= During flowering huge flower drop
- ☆ Pod borer resistant- ***Lablab pupurious lignosis***
- ☆ Soft poded indeterminate**- *L. parpurious typicus*** (Photosensitive)
- ☆ Determinate- Photo non sensitive
- ☆ Crop duration-
- ☆ Pole type = 210-240 days
- ☆ Bush type= 90-300 days
- ☆ Soil: Well drained loamy soil with pH range of 6.5-8.5.
- ☆ Rhizobial treatment: Treat the seeds with three packets (600 g) of ***rhizobiaum*** culture per ha using rice gruel as binder.
- ☆ Dry the treated seeds in shade for 15-30 minutes before sowing.
- ☆ Season: Bush type - Throughout year; Pole type – July-August.

- Dibble single seed 30 cm apart on one side of the ridge formed at a spacing of 60 cm for bush type. For pandal type, 2 - 3 seeds/pit at 2 x 3 m spacing. Spacing for CO-1 Dolichos bean is 1 x 1 m.
- Preparation of field: Plough the land to a fine tilth.Form ridges and furrows 60 cm apart for bush types.Dig pits of 30 cm x 30 cm x 30 cm at required spacing and fill it up with FYM and soil for pandal type.
- Irrigation: Immediately after sowing and on 3rd day, there after once a week.
- After cultivation: Hoe and weed thrice. Provide stakes to reach pandal of 2 m height and train the vines on pandal.
- Application of fertilizers: Basal dressing for bush type Manures and fertilizers Irrigated Dry FYM 12.5 t/ha 12.5 t/ha; N 25 kg/ha 12.5 kg/ha;P 50 kg/ha 25 kg/ha
- For pandal type Apply 10 kg FYM per pit (20 t/ha), 100 g of NPK 6: 12: 12 mixture as basal and 10 g N per pit after 30 days.
- Apply 2 kg each of *Azospirillium* and Phosphobacteria per ha at the time of sowing.
- Pests; Pod borer: Spray Carbaryl 50 WP thrice at fortnightly intervals @ of 2 g/l.Dust Carbaryl 10 D @ 25 kg/ha.
- Sucking pests: Spray Malathion 50 EC @ 1 ml/l or Dimethoate 30 EC @ 1 ml/l or Methyl demeton 25 EC @ 1 ml/l or Fenthion 1000 EC @ 1 ml/l at 15 days interval to control aphids and other sucking insects.
- Diseases
- Powdery mildew: Spray Wettable sulphur @ 2 g/l or Carbendazim @ 0.5 g/litre.

Cluster Beans: *Cyamopsis tetragonolobus* L; Fabaceae

- Origin- India
- Chromosome number 2n=14
- Ancestor of gour- ***C. senegalensis***
- Young plants contain hydrocyanic acid
- Flower colour- White/Pink
- Inflorescence- Raceme
- Gour gum extracted in- Cluster bean
- Short day plant
- 100 seeds=6gm
- Sandy loam soil is best
- Isolation distance-
- Foundation seed 50m
- Certified seed 25m
- It is hardy and drought tolerant crops
- Guar meal dry seed contain 33.3 per cent protein
- Green pods are rich in vit-A/iron
- Amide producing legume aspargine and glutamine
- Season- early crop- Feb-March
- Optimum temperature 23-33 ^{0}C
- Main crop June- July
- Green pod ready- to harvest 45 days
- Drought resist crop
- Two three productive irrigation is best
- Cluster bean seed contain 68-70 per cent galactoman non polysaccharide
- Harvesting- green pod 45 days maturity
- The lower powdery enough to start shattering
- Flower and pod development is critical stage of irrigation.
- Weed control-Fluchoralin- 2lit./ha
- Dry seed protein contains33 per cent
- Var.- Goma manjari- It is resistant to PM/bacterial wilt

- Pusa Mausmi- Rainy season
- Pusa Sadabahar- National var./both season
- Pusa navbahar- P. Mausmi × P. Sadabahar
- Sarad bahar- Self pollinated, selection
- P-28-1-1- Photo insensitive
- Spacing 45×15cm
- Giant type bean found in Gujrat
- It is also known as green manure crop
- Giant type- Highly guar green- grow in seed
- Dwarf type- pod type
- Rajasthan grown 82 per cent
- Mucilaginous galactomannan gum present in seed endosperm
- Seed rate 20-40 kg/ha
- Yield 6-8q/ha seed yield
- 60-80q/ha green pod
- Soil: Well drained sandy loam with pH range of 7.5-8.0. The crop tolerants salinity.
- Season and sowing: June - July and October - November. Dibble the seeds on the sides of the ridges 15 cm apart.
- Seed treatment: Treat the seeds with ***Rhizobium culture*** @ 600 g/ha using rice gruel as binder.
- Dry the treated seeds in shade for 15 – 30 minutes before sowing.
- Preparation of field: Plough the field to fine tilth and form ridges and furrows 45 cm apart.
- Irrigation: Irrigate the field immediately after sowing then at weekly intervals.
- Application of fertilizers: Apply FYM 25ton, ***Azospirillum*** @2 kg and Phosphobacteria @ 2 kg/ha, N 25, P 50 and K 25 kg/ha as basal and 25 kg N/ha 30 days after sowing.
- Pests; Leaf hoppers, aphids, and glasshouse whitefly Spray methyl demeton 25 EC or dimethoate 30 EC @ 1 ml/lit.
- Ash weevils: Spray phosalone 35 EC @ 1.5 ml/lit.
- Diseases; Leaf spot: Spray Mancozeb @ 2 g/lit.
- Powdery mildew: Spray Wettable sulphur @ 2 g//lit or dust Sulphur @ 25 kg/ha. Repeat it at 15 days interval.

French Bean: *Phaseolus vulgaris* L; Fabaceae

- Chromosome number- 2n=22
- Origin- Central America/Mexico
- Common name- Kidney bean, haricot bean, snap bean, navy bean, garden bean, string bean
- Seed are kidney shaped
- Shallow rooted crop
- 94 per cent pods, 1.7 per cent protein, 0.1 per cent CHO, 4.5 per cent fat, 1.8 per cent fibre dried bean are rich in protein and closely compare with meat
- Four cultivated species, progenatore- ***Phaseolus aborgineus***
- ***Phaseolus vulgaris***-Red kidney beans
- ***P. coccineus***- Scarlet runner bean
- ***P. lunatus***- lima bean
- ***P. acutifollius var. latifolius***- Moth bean
- ***P. coccinus*** are cross pollinated
- It is day neutral plants
- Semi pole var. -Short day plant
- Classification of cultivar given by Thomson and Kelly
- Optimum temperature range between 15-25 ^{0}C
- Crop is sensitive to frost high temp and high rainfall
- Soil- Acidic and Alkaline are not suitable
- Sowing season –
- 1st- July - September and many even extended upto October
- 2nd- January- February
- Hill- March- May
- Seed rate-
- Bush type- 65kg/ha (50-75 kg/ha.)
- Pole type- 25-30 kg/ha
- Flowering and pod development period are the critical stage
- Staking is an important operation for pole type
- Seed index can be used for maturity and seed fibre analysis

- Weed control- Fluchoralin-2litre/ha.
- Parachlorophynoxic acidic acid
- PGR- PCPA 2ppm alpha naphthyle acetamide
- Beta napthalic acetic acid 5-25ppm effect on fruit set
- Autogamous crop
- GA_3 50-200 ppm- Improve the crop growth
- Reduce vegetative growth CCC 400 ppm
- Tender crop and cascade- 1000 seed weight 250-600 gm
- Yield- 8-10 tonnes/ha bush type
- 12-15 tonnes/ha pole type
- Seed yield 1500-2000 kg/ha
- Crop ready 45 day after sowing bush type
- 70 days pole type
- Storage 5-7 °C, RH 80-85 per cent - 20-25 day
- Insect resistant- sel 2, SVM-1, Sel 9, sel4,
- Bush type 30×5 cm 250000 plant/ha
- Pole type 9×7.5 cm 80000 plant/ha
- 5BP7-mid tolerant genotype
- Insect pest- stem flies, thrips,
- Control- 2ml/lit. dimethoate
- Cornell 49/242 dominant gene(Are) for resistant to anthracnose
- Mites-Dicofol 2.5ml/lit.
- Serious disease- angular leaf spot- ***Phaeisariopis griseala***
- Disease- Yellow mosaic/anthracnose/rust/angular leaf spot/stem blight/ rhizoctonia root rot
- Control- Blitox or Difolition @2gm/lit.
- Root rot- capton 2gm/lit.
- Yellow mosaic- mono crotophus 1-25ml/lit.
- Rust- sulfex 2gm/lit.
- Low protein in French bean due to calcium/magnesium
- Bean common mosaic virus controlled by dominant gene (I)
- Toxic substance- Haemaglutine
- White seed var. contains more sugar

- ☆ Optimum soil temp for growth and yield 25-30 ^{0}C
- ☆ Physiological disorder
- ☆ Bean curl top virus (BCTV)
- ☆ Crop sensitive to high temp above 30 ^{0}C
- ☆ Blossom drops and ovule abortion high temp 35 ^{0}C is main
- ☆ Optimum temp for seed germination 12 ^{0}C below not tolerant cause
- ☆ Delayed flower initiation and development occurs below 10 ^{0}C
- ☆ Optimum temperature storage 5-7 ^{0}C
- ☆ **Transverse cotyledon cracking (TVCC)**
- ☆ Necrosis of hypocotyle- Low (Ca) content
- ☆ Non stringless var.- Long pods
- ☆ Variety-
- ☆ Pole type-
- ☆ Canadian wonder- Introduced from Canada
- ☆ Pusa Himlata- Early var., processing var., pureline selection
- ☆ Kentnicky wondere- Introduced from USA
- ☆ Lakshmi- torerant to angular leaf spot
- ☆ SVM 1- contender of (***P. vulgaris***)×(***P. multiflorus***), resistant to angular leaf spot
- ☆ Tweed Wonder, Blue Lake
- ☆ KKL1, YED1 hybrid between two pole var.
- ☆ **Bush type-**
- ☆ Pusa Parvati- Mutant var., early var., resistant to mosaic/PM
- ☆ Pant Anupam- Early variety, resistant to angular leaf spot
- ☆ Arka Komal- Very good for transport and good keeping quality, photo insentive
- ☆ Arka Suvidha- Photo insensitive, resistant to rust, multi resistant
- ☆ IIHR 220- Resistant to rust, pureline selection
- ☆ IIHR 909- Contender×Blue crop
- ☆ Jampa-Maxican var.- Highly resistant to yield
- ☆ Contender- early, tolerant to P.M/mosaic (USA)
- ☆ Bountyful (USA), Pant bean = (Turkish brown x Contender)
- ☆ Gaint stringless Sweden

- Potato bean-***Parchyrrhizus tubrosus***
- Arkaa Anoop- Resistant to bacterial wilt/rust photo insensitive
- Tender green bean aphid-***Aphis craccivora***
- Ooty one - Resistant to leaf spot/anthracnose pod borer
- Arka Sharath- Photo insensitive
- Pant Anupama -Resistant to angular leaf spot/bean common mosaic virus
- Soil: Well drained loamy soils with pH range of 5.5-6.0.
- Treat the seeds with Trichoderma viride @ 4 g/kg or Thiram or Carbendazim @ 2 g/kg ofseed 24 hours before sowing to control fungal diseases.
- In hills, sow the seeds in lines or in beds. In plains, sow the seeds on the sides of the ridges.
- Preparation of field: Hills: Dig the soil thoroughly and incorporate FYM. Form beds of convenient size.
- Plains: After two ploughings form ridges and furrows.
- Irrigation: Immediately after sowing, third day and thereafter once a week.
- Application of fertilizers: Apply FYM 25 t/ha at the last ploughing. N at 90 and P at 125 kg/ha should be applied on one side of the ridges
- For rainfed conditions of Shevar.oy hills, apply as a basal dose of 62.5 kg/ ha of Phosphorus as superphosphate and with another half of 62.5 kg/ ha Phosphorus as FYM enriched super phosphate.
- After cultivation: Weeding should be given 20 – 25 days and 40 – 45 days after sowing. The crop should be earthed up after each weeding.
- Pests; Whitefly: Place 20 yellow sticky traps coated with castor oil in polythene sheet to attract the white flies.
- Leaf hoppers, aphids, and glasshouse whitefly Spray methyl demeton 25 EC or dimethoate 30 EC @ 1 ml/lit.
- Ash weevils- Spray phosalone 35 EC @ 1.5 ml/lit.
- Pod borer: Spray Carbaryl 50 WP thrice at fortnightly intervals at @ 2 g/l. Dust Carbaryl 10 D @ 25 kg/ha.
- Diseases Powdery mildew: Spray Wettable sulphur @ 2 g/Lit. or dust sulphur @ 25 kg/ha.
- Rust: Dust Sulphur at 25 kg/ha.
- Anthracnose: Spray Mancozeb @ 2 g/l or Carbendazim @ 1 g/lit. or Chlorothalonil @ 2 g/ lit.

- Leaf spot: Spray Mancozeb @ 2 g/ lit.
- Root rot: Drench Carbendazim @ 1 g/lit.
- Mosaic: Select disease free planting material. Spray Dimethoate 30 EC @ 2 ml/lit. or Methyl Demeton @ 25 EC 2 ml/lit. thrice at fortnightly intervals.

Broad Beans: *Vicia faba* L; Fabaceae

- Chromosome number-2n=12
- Origin- Europe/Asia
- Bakla bean- Roundfali
- Faba bean/horse bean in a minor crop
- It is widely cultivated in latin America
- Cross pollination by insect
- Cluster of white black branches of flower
- Salt tolerant
- It is best for cool season
- Soil well drained loamy soil
- Soil pH 6.5-7.5
- Spacing 25cm
- Large seed 90-100 kg/ha
- Small seed 70-100 kg seed/ha
- N.P.K. 20: 50: 40
- Pod yield 7-10 tonnes/ha
- Green bean yield- 1.8-2 tonnes/ha
- Seed yield 2-3 tonnes/ha
- Var.- Pusa Sumit (improved var.)/Pusa Udit
- Ramba tall- Tolerant var.
- White seed var.- less soft
- Green seed- Soft and suitable for freezing
- Green pod caused -allergy favism- haemolytic anaemia
- Time of sowing- Oct.- Nov. rainy area
- High hill- March- April
- Long pod- Red epicure
- Mid hill – Sept.- Oct.
- Imperial white long pod
- Imperial green long pod
- Jawahar Vicia (MP) dormant type (mater piece green long pod)
- BR-2 (yellow seeded)

- ☆ Windor/small pod- BR-1 developed by bihar (black seeded)
- ☆ Imperial white Windsore
- ☆ Giant four seeded green windsore
- ☆ Temperature low as 4 ^{0}C
- ☆ Sulfer is also beneficial for production of broad bean
- ☆ This is only bean are grown as- wintercrop/and can survive at 4 ^{0}C
- ☆ Stem grown as like- sarvare shape
- ☆ Soil: Thrives in almost all soils with a pH range of 6.5-7.5 in hills only.
- ☆ Season and sowing: Sowe the seeds during July – August and November - December at 25 kg/ha at 45 x 15 cm spacing.
- ☆ Preparation of field: Plough the land to a fine tilth, level and form beds.
- ☆ Irrigation: Once in 5 days.
- ☆ Application of fertilizers: Apply 25 t of FYM and 50 kg P and 25 kg K/ha as basal dose. 25 kg N and 25 kg of K/ha are applied between 20 - 25 days after sowing and application of remaining 25 kg of N is done between 40 and 45 days.
- ☆ After cultivation: Earthing up is done on 45th day after sowing.
- ☆ As soon as the plants grow, flowering top is pinched off which causes the pods to develop early.
- ☆ Yield: 400 - 500 kg of beans/ha in 10 - 12 months.

Winged Bean; *Psophocarpus tetragonolobas*

- ✰ Origin-Africa (Madagascar)
- ✰ Chromosome number 2n= 22/18
- ✰ Madagascus in Africa, asparagus bean, multiflorus bean
- ✰ 20th century/God-scented vegetable
- ✰ It is very hardy plant
- ✰ Pods contain 1.2-2.9 per cent protein
- ✰ Leaf vegetable 6-15 per cent protein
- ✰ 25-32 °C best for better growth and development
- ✰ It is photo sensitive crop
- ✰ Evo cultivar.-
- ✰ Baqor, ribbon butterfly, mariposa
- ✰ Spacing120×60cm
- ✰ Lunita- flattant pod
- ✰ Short day plant and normal flower induction
- ✰ Dry matter yield will increase- phosphatic fertilizer
- ✰ Seed are sowing June-july
- ✰ Var.- IIHR- sel 20,60,71
- ✰ WB42- north east Parvatiya Anusandhan Kendra
- ✰ Alipasta- Double purpose, always year round var. (Meghalaya)
- ✰ Chimbux for seed production
- ✰ Timpanatise in animal is called by feeding of cluster bean
- ✰ Edible oil- 18 per cent
- ✰ 60-80 per cent manogalacton- guar gum present in
- ✰ Multipurpose- bean
- ✰ Anti oxidant to caferol
- ✰ Green pod 7-8t/ha
- ✰ Dry seed 1.5-2.0t/ha
- ✰ Tuber 0.6-1.2t/ha
- ✰ 18-32- flower production -short day
- ✰ 45-90q/ha- green pod
- ✰ 55-60q/ha- tuber

Lima Bean; *Phaseolus lunatus*

- ☆ Chromosome number- 2n=22
- ☆ Origin- Guatemala
- ☆ Fertile segmental allotetraploid 2n=44
- ☆ Sex form herkogamy
- ☆ Common name- Double bean, Butter bean
- ☆ Growth temp.- 15-20 ^{0}C dry cool climate
- ☆ When temperature given above 26 ^{0}C – Pod formation will reduce
- ☆ It is grown in twice in a year
- ☆ Rainy season June to July/2nd spring season
- ☆ Seed rate- 20-35kg/ha small type
- ☆ 45-50kg/ha large type
- ☆ Seed germination temp-15-30 ^{0}C
- ☆ Spacing 60×40 cm in pole type
- ☆ 35×25cm in semi pole type
- ☆ Seed are sown – eye down
- ☆ Large seeded variety adopt- lower temp
- ☆ It is also used to prepare from wine
- ☆ Tender pod ready 85-90 days
- ☆ Anderpetern has big (lima morphotype) large of late seed
- ☆ Yield- 5-10 tonnes/ha fresh green lima bean
- ☆ 5.4-10 tonnes/ha green shalled green
- ☆ Fertile segment auto tetraploid 2n=44 ***P.leutus×P. polystachyus***
- ☆ Pole type-
- ☆ Florida Butter/Challenger/King of Garden/Karollina Butter
- ☆ KKI-1- Butter bean
- ☆ **Semi pole type-**
- ☆ Hopi/Wilbur
- ☆ Bush type-
- ☆ Henderson Bush, Fordhook -242
- ☆ Baby potato/Baby Fordhook

- Downey mildew- ***Phytophthora phaseoli***
- Control- 0.2 per cent Dithan M-45
- Often cross-pollinated crops
- Pods storage 5-7 ^{0}C

Chapter 8

Tuber Crops

Sweet Potato: *Ipomoea batatas*

- ☆ Chromosome number- **2n=6x=90**
- ☆ Family- Convolvulaceae
- ☆ Origin- South America
- ☆ Proginator- ***Ipomea trifida***
- ☆ Also known sas- (White potato), (Irish potato)
- ☆ Carotene content- 5.4-20 mg/100gm
- ☆ Productivity is highest in Japan
- ☆ The genome- Ipomea has 650 species
- ☆ *Ipomea triloba* 2n=30
- ☆ Shape of tuber is fusiform
- ☆ 16 per cent starch having in S.P, 4 per cent sugar (20 per cent Alcohol producing material)
- ☆ Seeds are black in colour
- ☆ Short day plant
- ☆ Day length of 11 hrs is ideal for flowering
- ☆ Odisha/U.P./Bihar account for 89 per cent area, 88 per cent production
- ☆ Productivity of sweet potato in India is more than half of the world average
- ☆ 10 per cent starch, 3-6 per cent sugar, (good source of Vit-A)

- U.S.A about 60-70 per cent of the sweet potato is utilized for human food
- Sweet potato can be used as a source for used in the textile
- Sweet potato can be used as a high (CHO) feeding stuff for cattle/pig/poultry
- Sweet potato- Introduced- Europe
- Inflorescence- Cymose
- Flower colour- White/Pink
- Introduce in India- Portuguese in early 16 century
- Type of fruit- Capsule
- Anthesis-(4-5 am)
- Self incompatibility
- Majority of the pollen becomes dry by-(7am) and pollination become totally ineffective after (7am)
- It is the cheapest source of calories
- Cross pollination by- Bee/Bumble bee
- Day length of light intensity promote root development
- Ideal temperature for tuber formation 21-27 ^{0}C
- Best temperature for growth > 24 ^{0}C/less 20 ^{0}C
- Temperature below 10 ^{0}C plant will damage
- Sandy loam soil best
- Sweet potato- Acid tolerant plant pH 5.6-6.6
- Sensitive to- Alkaline and saline soil
- Propagation by- vine cutting
- Sweet potato is grown in India- Rainfed condition
- Spacing 30×60 cm Row×Row, 15×15 cm Plant×Plant
- General spacing 60×20 cm
- 82000 cutting required for one hac
- Storage after harvesting- 24-29 °C (80-90 RH) for 5-7 days
- 5-6 months for maturity
- Single harvesting and double harvesting are practiced in sweet potato
- Maturity of tubers gives a dark (greenish colour)
- Application of (K reduces) tuber cracking
- Fresh tuber- Yield 5-10 tonnes/ha

- In addition, 2.5-10 tonnes/ha for cattle feed
- Pest- sweet potato weevil (***Cylas formicarius***) tolerant variety – Rajendra sakarkand -47/5
- Control – Methyl demetron-(0.05 per cent)
- Phorate/carbofuron-(1.5kg/ha)
- New release variety Sree Arun, Sree Varun
- Nematode- ***Meloidogyne incognite***
- Shree Bhadra resistance variety
- Disease- sweet potato scab (***Spbacaeloma batatus***)
- Storage disease- soft rot/Rhizopus rot (***Rhizopus stolomifrus***)
- PGR- CCC/SADH best for increase tuber yield
- Weed control-
- Transgenic var.- jewel (moderate drought tolerant)
- Pusa Sundari- Rich in caroten
- Assam- Kalmegh- (very early var.)
- Bihar- Kalmegh×24, Sree Bhadra, RS-5, RS-35, RS43
- MH- Varsha- (rich in carotene 1200 IU/100mg
- Sree Bhadra, Pusa lal
- MP- kalmegh/Sree Bhadra -Excellent trap crop for root knot nematodes
- AP- Samrat/Kiran
- UP- VL/shakarkand
- Odisa- Gouri tolerant to mid season stress
- Sankar (excellent coocking quality)
- Rhuban
- Sree Nandini- superior cooking
- Jawahar sakarkand-114 (extra early var.)
- Konkan ashwani/Rajendra-35
- Deep purple colour (flesh colour creamy)
- Rajendra sakarkand (cross- 4×H-5), less prone to weevil
- Gold rush, Centennial, Velmon, Ranger
- Soil: Can be grown in loamy soil with a pH range of 5.6-6.6.
- Season and planting: June – July and September. Plant the terminal vine cuttings (82000/ha) at 20 cm spacing.

- ☆ The cuttings should be 10-15 cm in length with 2-3 nodes and to be collected from matured vines aged 3 months and above.
- ☆ Dip the vine cuttings in a solution by mixing 400 g of Azospirillum in sufficient quantity of water.
- ☆ Preparation of field: Plough the field to fine tilth.
- ☆ The soil depth should be atleast 30 cm. Form ridges and furrows 60 cm apart or beds.
- ☆ Irrigation: Irrigate before planting, on 3rd day and then after once a week.
- ☆ Stop irrigation one week before harvest.
- ☆ Application of fertilizers: Apply 25 t FYM/ha and 20: 40: 60 kg NPK/ha as basal and 20: 40: 60 kg NPK/ha after 30 days.
- ☆ If 20 kg/ha ***Azospirillum*** is applied, apply only 2/3rddose of N. It is preferable to apply N and P in the form of di-ammonium phosphate.
- ☆ After cultivation: The field should be kept clean by hand weeding till vines are fully developed.
- ☆ Earth up the field on 25th, 50th and 75th day after planting.
- ☆ The vines are lifted and turned on 50th and 75th day after planting but before earthing up to prevent root formation at nodes and to make the originally formed roots to increase in size.
- ☆ Spray Ethrel five times at 250 ppm at fortnightly intervals commencing from 15 days after planting.
- ☆ Sweet potato weevil:
- ☆ Remove previous sweet potato crop residues and alternate host i.e., Ipomoea sp. and destroy them.
- ☆ Use pest free planting materials. Dip the planting material in fenthion 100 EC or Monocrotophos 36 WSC @ 2 ml/l.
- ☆ Drench the soil with Endosulfan 35 EC or Fenthion 100 EC at 2 ml/l.
- ☆ Harvest the crop immediately after maturity and destroy the crop residues.
- ☆ Diseases; Soft rot: Spot drench with Carbendazim 1g/l.

Tapioca: *Manihot esculenta* Crantz.; Euphorbiaceae

- ☆ Native- Brazil
- ☆ Chromosome number 2n=36
- ☆ Common name- Topiaca/monioc/mandioca
- ☆ It is highly drought tolerant
- ☆ It is starchy root crop
- ☆ Nigeria is the major country gtowing 20 per cent of the world area
- ☆ 20 per cent world production in India
- ☆ It is monoecious spp
- ☆ Introduction in India- 1st time in kerla
- ☆ 50 per cent of area under cassava
- ☆ 98 spp of Manihot
- ☆ It is grown in rainfed crop and rainfed crop
- ☆ Flowering comes in 6 weeks after planting
- ☆ It is monoecious/producing male and female flower at the same inflorescense
- ☆ It is highly cross pollination- protoandry
- ☆ Propagation- stem cutting 15-30cm long
- ☆ Yellow colour of fresh is due to presence of carotene the precursor of vit- A
- ☆ Bitter and non bitter primarily due to bitter principle cynogenic glucoside HCN
- ☆ Female flower opens about 10 days before the male phase
- ☆ Fruit set and seed set indicate stigma is receptive from early morning 6: 00am
- ☆ Polyploid breeding method is most in cassava
- ☆ Tetraploid had been induce by treating 0.5 per cent colchicine for 24 hrs
- ☆ Soil- loamy soil is best, PH 5.5-7
- ☆ Major cassava grown soil
- ☆ Laterite- oxisol- black vertisole, pH 6.5-8
- ☆ Male sterility reported by byrne 1984
- ☆ Plant the setts vertically with buds pointing upward on the sides of ridges and furrows. 17,000 setts are needed for planting one ha.

- Cassava is drought tolerant crop
- Planting time- April- May in rainfed
- Irrigated: 75 x 75 cm (17,777 setts) and 90 x 90 cm (12,345 setts).
- Rainfed: 60 x 60 cm (27,777 setts).
- Under south conditions: 90 x 90 cm (12,345 setts).
- Thailand and India are the main exporter of cassava
- Starch in international market
- Sago prepared from cassava
- Cassava industries- located in solum-TN
- Ideal temp for cassava stem sprouting- 28-30 °C
- True cassava seed CTS developed in CTCRI
- **TCS seed rate- 1.5kg/ha**
- **Viability of cassava seed- 8 months**
- Cassava mosaic virus- CMV- is serious disease
- Transmitted by- white fly- ***Bemasia tabaci***
- Spacing- non branching 75-75cm
- Semi branched 90-90cm
- Crop ready 10-11 months
- Yield- 30-40t/ha- high yield var.
- 28-30t/ha- short duration var.
- Linamarine and losalstralin glucoside- present
- Responsible for rooting- ***Rhizopus/Bassillus spp./Penicillium***
- Invading pathogen are capable of releasing – polysaccharide and pectin degrading enzyme
- Var.-
- Shree Satysa- Multi hybrid 5 parent, release in 1977
- Shree Prakas- Released in 1987
- Shree Harsha- Starch 39.05 per cent
- Shree Jay– In 1998
- Shree Vijaya- Starch- 27-30- short duration
- Nidhi- 25t/ha
- KMC1- 31.4 per cent Starch
- Shree Vishakam- Hybrid var.

- ☆ Shree padamnabha- CMDA- Resist. Cassava var.
- ☆ Cynogenic potential pf cassava leave is 15-20 time higher than that of edible part
- ☆ Enzyme responsible for hydrolysis are lima marinase
- ☆ Cassava contains 70 per cent ethanol/soluble sugar 1-6 per cent
- ☆ Gari/fasiral/fulgar/mocroni- Prepared from cassava
- ☆ Resistant to casava mosaic virus- CO-3
- ☆ Tolerant to cassava mosaic virus- H-97
- ☆ Triploid var. Shree Harsha 2n=3x=54
- ☆ Resist- cercospora leaf spot and tolerant drought- Shree Prakash
- ☆ H-7/H-226/- soya production var. 0 starch 28-30 per cent
- ☆ CTCRI CO (Tp) 5 (Sree Padmanabha) Resistance to cassava mosaic disease.
- ☆ Season June – July; October – November. Fairly good starch content (28 per cent). Moderate tuber yield (38 t/ha).
- ☆ Erect and top branching habit.
- ☆ Regular flowering and seed setting.
- ☆ Tubers long, cylindrical, white flesh and low in cyanoglucoside. Duration 270 – 300 days.
- ☆ Thrives best in tropical, warm humid climate with well distributed rainfall of over 100 cm per annum.
- ☆ Cultivated up to an elevation of 1000m.
- ☆ Season and planting: Plant throughout the year under irrigation.
- ☆ Select healthy mosaic free vigorous plants for taking planting materials.
- ☆ Prepare setts of 15 cm long with 8 – 10 nodes from the middle portion of the stem.
- ☆ Avoid mechanical damage while preparation and handling of setts.
- ☆ The cut end should be uniform. Dip the setts in Carbendazim 1g in one l of water for 15minutes before planting.
- ☆ Preparation of field: Plough the field 4 – 5 times to get a fine tilth.
- ☆ The soil depth should be at least 30 cm. Form ridges and furrows at the following spacing:
- ☆ Irrigation: First irrigation is given at the time of planting.
- ☆ Life irrigation is given on the 3rd day followed by once in 7 – 10 days up to 3rdmonth and once in 20 – 30 days up to 8thmonth.

- ☆ Drip irrigation: Install drip system with main and sub-main and place the inline laterals at the interval of 1.5 m.
- ☆ Place the drippers at the interval of 60 cm for 4 LPH and 50 cm for 3.5 LPH in the lateral system.
- ☆ Form the raised beds at 120 cm width at an interval of 30cm and place the laterals at the centre of each bed.
- ☆ Sequential cropping: Raise CO 2 vegetable cowpea during March and harvest green pods before planting cassava during June –July.
- ☆ After green pod harvest, incorporate the cowpea haulms into the field by disc ploughing. Through sequential cropping, 50 per cent reduction in application of FYM (12.5 t/ha) and P (30 kg/ha) is ensured.
- ☆ Manuring; Irrigated crop: Apply 25 t FYM/ha and incorporate at the time of planting. Apply 45: 90: 120 kg NPK/ha as basal and 45: 120 kg NK/ha 90 days after planting during earthing up.
- ☆ Rainfed crop: Apply FYM at 12.5 t/ha along with 50 kg N, 65 kg P and 125 kg K/ha as basal. Apply 2 kg of Azatobactor through soil application at 30 – 60 days after planting on receipt of showers (2.0 kg Azatobacter + 20 kg FYM + 20 kg soil per hectare).
- ☆ Chlorosis: Foliar spray of 1 per cent FeSO4 + 0.5 per cent ZnSO4 at 60 and 90 DAP.
- ☆ After cultivation: Fill up the gaps within 20 days of planting.
- ☆ Carry out 1st weeding 20 days after planting.
- ☆ Subsequent weedings should be done once in a month upto 5 months depending upon the weed intensity.
- ☆ Thin to two shoots per plant during 60th day.
- ☆ Grow aggregatum onion, coriander, short duration pulses and short duration vegetables as intercrops from planting date up to 60 days
- ☆ Pests; Mites: Soaking spray with dicofol 18.5 EC 2.5 ml/l during 3rd and 5th month.
- ☆ White fly (*Bemisia tabaci*): Spray neem oil 3 per cent or fish oil rosin soap 25g/l
- ☆ Apply Methyl demeton in the early stage and Phosalone in the later stages of crop growth. Spiralling whitefly: Spray Dichlorvos 76 WSC @ 1 ml/l or Triazophos 40 EC 2 ml/l. Add wetting agent.
- ☆ Diseases; Mosaic: Select the planting materials from healthy plants.
- ☆ For the control of white fly vectors, adopt IPM practices mentioned above.

- Cercospora Leaf spot: Spray Mancozeb at 2 g/l twice at 15 days interval.
- Tuber rot: Avoid water stagnation. Give good drainage facilities.
- Spot drench with Copper oxychloride 2.5 g/l or apply through soil ***Trichoderma viride*** @ 2.5 kg/ha as basal and at 3rd and 6th month after planting
- Crop duration: 9 to 11 months.

Elephant Foot Yam: *Amorphophallus campanulatus* Blume; Syn: *A. paeoniifoliu*, F-Araceae

- ☆ Chromosome number 2n=28
- ☆ Origin- South -East Asia (India and Sri Lanka)
- ☆ Inflorescence- Spadix
- ☆ Spacing 90×90cm
- ☆ ***Animophyllus compenulatus/paeonifolious***
- ☆ ***Amarophophallus paeonifalius of hortansis***- Cultivated and edible
- ☆ ***Amarophophallus of sylvertis***- Wild with rough petiole
- ☆ Acts as blood purifier
- ☆ Anti nutrient factor present in taro is calcium oxilate
- ☆ Propagation- stem cutting and tuber
- ☆ Weight of tuber piece for planting 500-1000gm
- ☆ Dormancy 5-6 months, broken by thio urea (0.1 per cent GA_3, ethral)
- ☆ None synchronization of flowering extreme (protandry result) in cracking of seed set
- ☆ Gajendra, Santragachi, Shree padma, Bidhan kusum, Shree athira,
- ☆ Pedigree (Shree Padma× AM-45)- Hybrid
- ☆ Soil and climate: Rich red-loamy soil with a pH range of 5.5-7.0 is preferred.
- ☆ It is a tropical and subtropical crop. It thrives well with a mean annual temperature of 30-35 ^{0}C.
- ☆ It requires well distributed rainfall of 1000-1500 mm spread over a period of 6-8 months, with humid and warm weather during vegetative phase and with cool and dry weather during the corm development period.
- ☆ Season and planting: It undergoes dormancy period of 45 to 60 days.
- ☆ Traditionally farmers take advantage of the dormancy period by planting during February-March so that the setts would sprout with the pre-monsoon showers. April – May is the planting season.
- ☆ The tuber is cut into 750-1000g small bits in such a way that each bit has atleast a small portion of the ring around each bud.
- ☆ Whole corms of 500 g size can also be used as a planting material.
- ☆ Use of cormels and mini-sett transplants of 100 g size as planting material at a closer spacing of 45 x 30 cm is also suggested.

- There are also projections with tender buds called "Arumbu". These are removed before planting as they do not give vigorous growth.
- An ordinary sized yam gives about 6 to 8 bits for planting.
- The cut pieces are dipped in cow dung solution to prevent evaporation of moisture from cut surface.
- In some places, the small round daughter corms are also planted.
- The cut pieces are planted in beds at 45 cm x 90 cm spacing or pit of 60 x 60 x 45 cm size is dug and planted.
- The pit should be filled with top soil and farm yard manure (2kg/pit) prior to planting.
- The pieces are planted in such a way that the sprouting region (the ring) is kept above the soil. About 3500 kg of corms will be required to plant one hectare.
- Sprouting takes place in about a month.
- Preparation of field: The land is brought to fine tilth and beds of convenient size are formed.
- Intercropping: Vegetable cowpea var. CO-2 is recommended as suitable intercrop in elephant foot yam.
- It can be intercropped profitably in coconut, arecanut, rubber, banana and robusta coffee plantations at a spacing of 90 x 90 cm.
- Half quantity of FYM (12.5 t/ha) and one third of NPK (27: 20: 33) will be sufficient for the intercrop.
- Irrigation: It is mostly raised as a rainfed crop.
- Water stagnation is harmful to the crop.
- Irrigation facility is available, irrigation can be given once a week.
- Mulching immediately after planting not only conserves soil moisture and regulates soil temperature but also suppresses weed growth.
- After cultivation: Weeding and earthing up as and when necessary.
- Collar rot: The disease is caused by a soil borne fungus ***Schlerotium rolfsii.***
- Water logging, poor drainage and mechanical injury at collar region favour the disease incidence.
- Harvesting: It becomes ready for harvest in about 8-9 months after planting.
- The crop attains maturity when total senescence takes place.
- Yield: 30 – 35 t/ha in 240 days.
- For seed purpose, the yams can be left in the field itself till planting the next crop or the lifted yams can be stored in sand or paddy straw.

Taro: *Colocasia esculenta* L. Scott; Araceae

- ☆ Origin- India and Southern Asia
- ☆ Chromosome number- 2n=2x=28, 3x=42
- ☆ Colocasia name derived from- Egyptian word
- ☆ Taro recomonded for gastric patients
- ☆ Taro flour is good for baby food
- ☆ Taro corm are used for fermented acidic product
- ☆ Miestubre disorder of Colocasia is due to calcium deficiency
- ☆ Starch is also used for preparation of baby food
- ☆ Propagation corm and carmel
- ☆ Inflorescence-Spadix
- ☆ Edoe- ***Colocasia esculenta var. antiquarum*** widely grown in India
- ☆ Edoe type produce 15-20 carmel
- ☆ Dasheem- ***C. esculenta var. esculenta***= 4-5 cormel produce
- ☆ Weight of small cormel 20-25gm
- ☆ pollination by insect
- ☆ 17-25 per cent amylose starch found in taro corm
- ☆ 13 spp found in Colocasia group
- ☆ Spacing 60×45cm
- ☆ Sree Resmi, Shree Pallavi, Bidhan Chaitanya, Bidhan Jayleb
- ☆ Panchmukhi- Fived faced
- ☆ Sahasramukhi, Khasi Bhunga, C-9
- ☆ Taro is vegetatively propagated and 1cm corm tip
- ☆ 15-20 cm petiole called huli are used
- ☆ The leaves, petiole, corms and cormel are edible
- ☆ Soil and climate: It come up well in loamy soils with a pH range of 5.5-7.0 and a combination of warm and moist climate with a mean temperature of 21-27 ^{0}C.
- ☆ It can be grown up to 1500 m elevation with well distributed rainfall of about 1000 mm during growth period.
- ☆ In areas where rainfall is less, a good amount of supplementary irrigation is required for successful production.
- ☆ Planting material and seed rate: Cormels weighing about 20-25g form good planting material.

- ☆ Seed rate of 800 kg/ha is required.
- ☆ Season: Under rainfed condition, planting during April to June is ideal. June-July and February – March. If grown as irrigated crop, it can be raised throughout the year.
- ☆ Preparation of field: Plough the field to a fine tilth and form ridges and furrows at a spacing of 45 cm. In sandy loam soil, pit method is followed.
- ☆ Plant at a spacing of 45 cm in furrows. The cormels may be planted to a depth of 2.5 to 7.5 cm.
- ☆ Mulching: Planted seed tubers take 30 to 45 days for sprouting. Mulching helps to hasten sprouting and control weed growth.
- ☆ Gap filling: Under field conditions, 5-10 per cent of the seeds fail to sprout.
- ☆ To overcome this situation, about 2000-3000 corms/cormels per hectare may be planted in a nursery at a close spacing so that sprouted tubers from the nursery can be used for gap filling.
- ☆ Irrigation: Once in a week.
- ☆ After cultivation: Weeding and earthing up should be done 45 to 60 days after planting.
- ☆ Deep cultivation should be avoided.
- ☆ Small inefficient suckers from the mother plants have to be removed along with the second weeding.
- ☆ It requires profuse irrigation and shade.
- ☆ Crop protection:
- ☆ Aphids and worms are pests attacking the leaves.
- ☆ Control: Spray Quinalphos or Dimethoate 0.05 per cent. Mealy bugs and scale insects damage cormels and corms and hence select cormels free of these pests for planting.
- ☆ If infested, the seed cormels should be dipped in Dimethoate or Monocrophos 0.05 per cent solution for 10 minutes.
- ☆ Leaf blight (***Phytophthora colocasiae***): Oval or irregular purplish or brownish necrotic lesions with watersoaked periphery appear on leaves.
- ☆ Heavy incidence causes up to 50 per cent crop loss.
- ☆ Management: Use of field resistant Varieties *viz.*, Muktakeshi and Jankhri, spray with fungicides *viz.*, Mancozeb (0.2 per cent) or Ridomil MZ 72 @ 2g/l of water Harvesting: Crop will be ready for harvest in 6-8 months after planting.
- ☆ Yield: 8 – 10 t/ha in 180 days.

Dioscorea: *Dioscorea alata* and L.; *Dioscorea esculenta* L. Dioscoreaceae

- Female flower has inferior ovary
- *D. ratundata*- extensively cultivated in Africa
- Temperature 20-30 ^{0}C promotes better growth
- The main sapagenin present in diosgenin
- Which is the starking point of several cardicosteriod drugs
- Poisonous type- ***D. durentorum*, *D. hispida***
- Africa alone produce 90 tuber and cover 85 per cent of area yam contain 18-20 per cent starch amylopectine
- Fufu is an important product made from yam
- Yam flavour is also used for human consumption as kokonte in northern zone of west Africa
- *D. alata* is used for making wafer chips and biscuit
- *Dioscorea* genus is dioecious in nature
- Fruit type trilocular capsule
- *D. alata* no in seed dormancy exist
- Amylose content range from 10-25 per cent
- Lesser yam ***D. esculenta*** is sweet and mature easily than other yams
- Propagation by tuber species/small whole tuber
- Tuber top portion is known as proximal
- Tuber middle and bottom portion is known as distal
- Tuber weight of propagation lesser yam 125-150gm
- Greater yam 200-250 gm
- It requires warm and humid conditions with a mean temperature of 30^{0}Cand a well distributed annual rainfall of 1200-2000 mm.
- Sandy loam soil with a pH of 6.0 to 6.5 is preferred with good drainage and cool weather.
- Season and planting: March - June.
- 8-9 month harvesting
- Vine cutting- slow tuber production
- Long dormancy period 2-3 months
- Dormancy breaker- ethylene chlorohydrazine followed by dry storage

- ☆ Soaking of tuber in 1 per cent MH delayed the sprouting storage
- ☆ 30gm mini sett is best for rapid seed yam production
- ☆ Peste – scale insect- as pidiella horthii
- ☆ Double harvesting is done in yam
- ☆ Male/female flower are borne on separate plant
- ☆ Male flower area on (long axillary panicle)
- ☆ Female borne – Exillary spike
- ☆ Spacing 75×75cm ***D. esculenta***
- ☆ 90×90cm ***D. elalata*, *D. rotundtata***
- ☆ Yam generally 30-40 days for sprouting and another 2 months to develop a vegetative cover on th ground
- ☆ several dwarf and semi dwarf progeny ***D. ratudata***
- ☆ Chinese yam/lesser yam**=** ***D. esculenta***
- ☆ Hard yam/pacific yam**=** ***D. nammularia***
- ☆ African yam/white yam=***D. ratundata***
- ☆ Chinomom yam**=** ***D. apposite***
- ☆ Yellow yam- ***D. canynsis***
- ☆ Potato- ***D. bulbifera***
- ☆ *D. alata*-
- ☆ Var.- greater yam
- ☆ Higher yield- Shree Keerthi, Shree Rupa
- ☆ CTCRI- Shree Shilpa- F1 hybrid, Shree Kartika, Konakan Gorkand
- ☆ ***D. esculenta*—**
- ☆ Lesser yam- high yield
- ☆ Var.- Shree Latha- 1st edible cv. Shree Kala
- ☆ CTCRI- Konakan Kanchan
- ☆ ***D. ratundata-***
- ☆ White yam
- ☆ Var.- Shree Subra, Shree Priya
- ☆ CTCRI- Shree Dhanya- Dwarf var.
- ☆ Tuber crop- heavy- Potassium feeder
- ☆ ***D. spp*.** produce areal tuber or bublite in leaf axil**-** ***D. alata***
- ☆ World 1st hybrid of ***D. alata***- Shree Shilpa

- ☆ Dwarf var. of ***D. ratundata***- Does not need staking- var.- Shree Dhanya
- ☆ Minimum tuber weight of ***D. alalat*** common *D.* ***ratundata***= 250gm
- ☆ Yellow colour spp- ***D. corensis***
- ☆ Single tuber of minimum- 100gm for commercial cultivation of ***D. esculenta***- 100gm
- ☆ Female greater yam- Shree Rupa
- ☆ Ovule shape- solen tuber- Shree Shilpa
- ☆ Greater yam with a digitate tuber and resist to- scale ***insrectishreerupa***
- ☆ Branched steam var. of greater yam- Shree Shilpa
- ☆ Shree Priya- seed selection- umidika
- ☆ Ahree Shubra seedling section
- ☆ Shree Dhanya- seedling selection of siv progency
- ☆ Male right yam by CTCRI- Shree Dhanya
- ☆ Maximum contribution in world- *D. rotundata*
- ☆ Best time of pollination 1 am
- ☆ Fruit is capsule, seed is winged
- ☆ Hybridization easy in – white yam
- ☆ Number of spp for edible 12
- ☆ Use mature tubers or pieces of 250-300 g tubers taken from the previous crop as seed material at the rate of 1875 – 2500 kg/ha.
- ☆ Mini setts of 25 g are recommended for planting directly in the field or raising a nursery and planting plants after 60 days.
- ☆ For planting lesser yam, medium sized tuber of 100-150 g is sufficient.
- ☆ Planting is done in beds or in ridges or in mounds or in rows 75 cm apart either way.
- ☆ Preparation of field: Plough the field to a fine tilth and form ridges and furrows at 75 cm spacing for raising lesser yam.
- ☆ Ridges and furrows at 75 cm spacing or pits of 45 x 45 x 45 cm for planting greater yam at 90 x 90 cm. Fill the pits with top soil and FYM.
- ☆ Method of planting: To plant greater yam, 3000-3700 kg and for lesser yam 1800-2700 kg of seed material is required.
- ☆ Irrigation: Copious watering once in a week is necessary.
- ☆ Application of fertilizers: Apply FYM @ 25 t/ha at the time of last ploughing.

- ☆ Follow fertilizer schedule of 40: 60: 120 kg NPK/ha as basal and 4 kg/ha of Azospirillum (mixed with 40 kg of soil) 30 days after planting.
- ☆ Apply 50 kg N and 120 kg K/ha 90 days after planting.
- ☆ It can be intercropped profitably in coconut, arecanut, rubber, banana and robusta coffee plantations at a spacing of 90 x 90 cm.
- ☆ Trailing: Trailing is necessary to expose the leaves to sunlight.
- ☆ It is done within 15 days after sprouting by coir rope attached to artificial supports in the open area or to the trees where it is raised as an intercrop.
- ☆ Crop protection: Yam scale is found to occur on the tubers both under field and storage conditions. As a prophylactic measure, dip the planting material in Monocrotophos 0.05 per cent.
- ☆ Harvesting: Greater yam and white yam become ready for harvest by 9-10 months after planting. Lesser yam takes 8-9 months for attaining maturity.
- ☆ Yield: 20 – 25 t/ha in 240 days.

Chapter 9

Underutilized Vegetables

Chinese Potato: *Coleus parviflorus* L.; Syn: *Solenostemon rotundifolius*; Labiatae

- ☆ Varieties: CO-1, Sree Dhara.
- ☆ Soil and climate: Red, loamy and well drained soils.
- ☆ Comes up well in shade with warm humid climate.
- ☆ It requires very good rainfall for its growth and cannot withstand drought conditions.
- ☆ In case rains are not received, irrigation has to be provided for satisfactory growth.
- ☆ Raising nursery: Raise a nursery, approximately one and half months prior to planting.
- ☆ An area of 500 m^2 is required to produce vines for planting one hectare of land.
- ☆ Cattle manure or compost may be applied @ 1 kg/m^2 and ridges/mounds may be prepared at a closer spacing of 45/60 cm.
- ☆ Healthy tubers that weigh about 15-20 g may be planted at 5cm spacing on the ridges/mounds so as to accommodate 750-1200 kg tubers in 500 m^2 area during March- April.
- ☆ Top-dress with urea (5 kg/500 m^2) at about three weeks after planting to encourage good vine growth.
- ☆ Stem cutting of 15-20cm length from these sprouts also used as planting material.

- ☆ To enable rapid multiplication of the planting material, single node cuttings can be planted directly in the secondary nursery.
- ☆ Such single node cuttings produce axillary shoots within one week can be planted in the main field.
- ☆ Preparation of main field and planting: The field is ploughed 4 to 5 times to a fine tilth and form ridges and furrows 60 cm apart.
- ☆ Use herbaceous cuttings of 10 cm length taken from the nursery beds and plant in the main field during July – October at a spacing of 30 cm on the ridges either in vertical or horizontal position.
- ☆ Horizontal planting of vines to a depth of 4-5 cm and exposing the terminal bud ensures quick establishment and promote tuber yield.
- ☆ Irrigation: At weekly intervals.
- ☆ Application of fertilizers: FYM at 25 t/ha as basal and NPK at 30: 60: 150 kg/ha. Apply 30 kg N 30 days after planting at the time of earthing up along with 2 kg of Azospirillum.
- ☆ Cultivation of sweet potato (Sree Bhadra) as a preceding crop in May-June enables trapping of root knot nematodes in the soil.
- ☆ To control leaf folding caterpillars and vine borers, dipping the vines in insecticide solution (Dimethoate or Rogar 30 EC ie. 1.7 ml/litre) for 10 minutes prior to planting is helpful.
- ☆ Harvesting: Harvest the crop when the vines dry up at 4-5 months after planting.
- ☆ Yield: 15 to 20 t/ha in 120 days.

Horse Radish; *Armoracia rusticana*; F. Cruciferae

- ☆ Origin-East Europe
- ☆ Chromosome number=2n= 4x= 32, n= 8,
- ☆ Cylindrical root and Perennial root crops
- ☆ Pungency is due to isothiocynate
- ☆ Two type-
- ☆ Common type – broad crinkle leaf and root of high quality.
- ☆ Bohemian type= narrow leaf and poor quality
- ☆ Leaf used salad and sandwiches
- ☆ Propagation by side root and crown

Globe Artichoke; *Cynara scolymus*

- ☆ Family- Compositeae
- ☆ Origin- Mediterranean region
- ☆ Chromosome number 2n=34
- ☆ Well known vegetable in Europe
- ☆ Common name-Globe artichoke/French artichoke/good gifted veg.
- ☆ Perennial herbs 1-15m height
- ☆ Edible part- Immature flower bud/globular immature
- ☆ Thickend bracts phyllaries bud
- ☆ Heat receptacle
- ☆ Choke- flower buds are eaten
- ☆ Flowering- Feb- May in northern India
- ☆ Small heads are eaten only after cooking
- ☆ Artichoke derived- Arabic word- barbs of the earth
- ☆ Thick heart is used for canning
- ☆ Active ingredient -cynorin
- ☆ Use food of dietary purpose (diabities)
- ☆ Leaves are bitter and considered to be useful in dropsy and rheumatism
- ☆ The shoot produces a rosette of deeply
- ☆ Cut pin notified, wooly leaves 5-8cm long
- ☆ Flower heads are considers to be useful in dietary of diabetes
- ☆ New growth emerge in spring remove by off-shoot formation (root stock)
- ☆ Distance row 120×90cm
- ☆ It is a cool season crop 12-18 ^{0}C growing temp.
- ☆ Temp. below 28°C cause tough and tender to spread occur
- ☆ 1 ha= 6900 plant
- ☆ In Uttarakhand green globe plant could be harvested 225 days after planting
- ☆ Freezing temp. (-2 ^{0}C-1 ^{0}C) prior to harvest kills the bud and above ground growth
- ☆ 3-5 irrigation required
- ☆ Seed sowing and planting- Broadcasting method August- October

- ☆ Propagation by seed (less polular) inferior quality
- ☆ By sucker commercial used
- ☆ In hills March to May
- ☆ Seedling transplant 10-12cm height
- ☆ Sucker and upshoot are removed when they are 30-45 cm in height spacing 240×180cm in winter California
- ☆ Plant grown from seed come to bearing in about 8 months
- ☆ Plant grown from sucker comes to bearing in about 6-7 months
- ☆ Harvest season Sept-Oct is best
- ☆ RDF nitrogen 120-140kg/ha, P_2O_5 100 kg/ha, K_2O 400-500kg/ha
- ☆ Globe artichoke head range from 2.5-10 cm diameter
- ☆ Cool season crop temperature 12-18 ^{0}C
- ☆ Cool season is best for crop rise
- ☆ A mature plant produces 12 or more stem
- ☆ 45-50 edible buds
- ☆ Buds should be harvest at 2.5-3.0 m stem length
- ☆ Storage- 0 ^{0}C and 90-95RH for 3-4 week
- ☆ Sowing time low hills June July, Aug- Oct
- ☆ Mid hills March – May
- ☆ Leaves colour blue green sometime spiny
- ☆ Progenatore globe artichoke, cordoon (***Cynara carduncul**us*)
- ☆ Perennial planting produce for 3-7 year
- ☆ Mature plant grown about 1.5 m
- ☆ Seedling transplanted 10-12cm height
- ☆ Flower bud contain insulin
- ☆ Pre cooling of bud at 5 ^{0}C is practiced to achieve less/weight loss/ discolourtaion/incidence of decay
- ☆ Yield/plant 475gm and mean yield/m2 1.55kg respectively
- ☆ Yield 10-12tonnes/ha
- ☆ Cross pollination by insect
- ☆ Variety- green globe/purple globe- are popular var.
- ☆ Violet de province fresh market
- ☆ Catanese- fresh market

- Spinosa- fresh market
- Spinosa sarda, biancotarantino-for processing
- Produce de jesi- processing
- Ball/tudalla/E15/brindisino/Romanesco are poular variety
- Yield 10-12 tonnes/ha
- Light frost may not destroy the edible quality of bud but cause blistered appearance
- Disease-
- Botrytis rot- botrytis cinerea in rainy season, methaloxy 10.025 per cent
- Tri demofone o.5gm/lit.
- Physiological disorder-
- Scuffing
- Split bracts
- Detached epidermis
- Bract wilting – All are marketing disorder
- Bacterial crown rot
- Curly dwarf

Chekurmanis; *Sauropus androgynous*

- Origin- Indo-Burma
- Edible parts green leaves
- Its is also known as vegetable of 21st century
- It is commoly called multivitamin green
- Due to high nutritive value
- Propagation by stem cutting, soft wood and semi hard wood
- 500ppm IBA/IAA treated before planting
- It is a perennial/small shruby leaf veg. show growing
- It is 1st introduced into Kerala from Malaysia 1953 and known as malyalam as maldurakheera
- Inflorescences is axillary with small radish flower
- Flowers are monoecious
- The crop is very highly cross pollinated and entomophilous
- Because of protogynous and monoecious nature of flower
- The leaves are very rich in protein 6.8-7 per cent as compared to smsrsnthud 3.2 per cent
- Training- bowers system
- Planting- 100000 cutting/ha
- Annual productivity 25-45ton/ha
- Per plant yield 1-3kg leaves/year
- 1st succulent leaves may be done after 3-4 months

Rhubarb; *Rheum rhaponticum*

- ☆ Chromosome number-2n=4x=44
- ☆ Native- Temperate Asia
- ☆ Origin- Siberia
- ☆ Polygonaceae
- ☆ Sex form-protandrous
- ☆ It was 1st grown in Italy
- ☆ Economic part-Petiole
- ☆ Propagation by seed it is not recommended
- ☆ They are used for sances and pies
- ☆ Propagated by the division of crown
- ☆ Plant is herbaceous perennial
- ☆ Crown are planted during spring
- ☆ Fibrus root system
- ☆ Disorder- pithiness/splitting
- ☆ 18th century introduced in USA
- ☆ Flower colour small greenish- White flower
- ☆ Self pollination
- ☆ Snout beetle damage- Plant
- ☆ Lower flowers are fertilized by pollen from upper flower
- ☆ Foot rot and crown rot in serious disease- fungus
- ☆ Oldest var.- Victoria/Linnaeus
- ☆ New var.- red stalk- McDonald/ruby/veluntine/sunrise/strawberry/cherry red/Victoria
- ☆ Climate- Cold resistant plant
- ☆ Optimum temp for proper development 15.6 °C critical low temp would be 3.3 °C below which the vegetative part damage
- ☆ Production of rhubarb stalk during the winter forcing of rhumbarb
- ☆ Temp. goes above 23.9 °C during summer
- ☆ 10 °C growth will be poor

- Well drained loamy soil is best
- Harvesting from 2nd year of planting
- It is tolerant soil acidity
- High temp. the crown sprout and the stalk grown out

Yam Bean; *Pachyrhizus erosus*

- ☆ Family: Fabaceae
- ☆ Chromosome number 2n=22
- ☆ South America
- ☆ Mishrikand/potato bean/Mexican yam bean
- ☆ Edible part tuberous root
- ☆ Also called as sakalu
- ☆ Propagation by seed
- ☆ Seed rate 20-60 kg
- ☆ Spray of 2,4-D (50ppm) flower initiation stage increase the tuber yield
- ☆ Harvesting 110-140DAS, markets demand at a time of Sarswati Pooja.
- ☆ Harvesting delay tuber cracking
- ☆ Potassium reduces the tuber cracking
- ☆ Mexican type yam bean – no preferred in India, large size of tuber, light sweet and prone to tuber cracking
- ☆ Local type yam bean- preferred in India, small size, of tuber, more sweeter, white flesh and not prone to tuber cracking
- ☆ Rajendra mishrikand-1 popular in Bihar yield is 3-40 t/ha.

Tania; *Xanthosoma sagittifolium*

- Family: Araceae
- Chromosome number 2n=2x= 26
- West Indies in central America
- Edible part leave, tuber
- There are no specific released variety of tania in India
- It is an herbaceous perennial species of tropical flowering plants
- Inflorescences –Spadix, spathe
- Growth cycle last from 9-11 month during the first six month the corms and leaves develop in the last four month
- This perennial can grow up to 270cm tall.
- Temperature ranging from 13-29 ^{0}C crops suited high rainfall 140-200cm. pH 5.5 to 6.5
- Tania grown in backyard (kitchen garden) in India
- Tania is grown throughout the year
- Planting material most commonly used are portion of the central corm (100-150gm with three to four buds)
- One month old sprouted plants raised in nurseries can also be used.
- A spacing of 60cm x 60 cm or 75cm x 75cm can be adopted
- The first six months is a critical period for weed control.
- The most serious disease leaf burning disease and root rot disease.
- Harvesting of the crops is usually done during ninth to eleventh month after planting.
- Local variety are grown in India (low acridity)
- It is closely related to colocasia (medicinal value and)
- Propagated by- corms and cormels
- Grown as a ratoonig crop for 5-6 year.
- Spacing 60 x 60 cm
- Potato bean can not stand in waterlogged condition
- For corms crops should be left in the soil up to 6-9 month without deterioration
- Harvested corms should be cured in sunlight for 4-5 days
- It gives yield of 500-600q/ha

Chapter 10

Okra (Lady Finger/Bhindi); *Abulmoschus esculantus*

- Origin—Tropical Africa
- Sanskrit name tindisha, gnadhmula, karanparn phal
- Chromosome number 2n=2x=130
- Okra is a warm season crop
- Rich source of Iodine
- No. of seeds 17 seed/gm
- Type of fruit: capsule
- Rich source of calcium (66mg), potassium 103 mg/gm, vit- C
- Fiber present in okra 1.2 gm/100gm iron 1.50gm/100 gm
- Mucilage present in okra fruit is (polysaccharides) galacturonic/glucoronic acid
- Oil contains in okra seed 13-22 per cent
- Protein contains in dry okra seeds 20-24 per cent
- Root used for sugarcane juice cleaning
- India is leading producing country 70 per cent production
- Leading okra producing state in India W.B, Odisha
- Highest productivity of Okra (A.P, JH, BH)

- ☆ After onion okra account 70 per cent of 30 per cent exchange earning from export of vegetable
- ☆ Genus *Abelmoschus* was 1st established by Medikus in 1787
- ☆ No inbreeding depression,
- ☆ Seed multiplication ratio 1: 100
- ☆ Cultivated okra is polyploidy in nature
- ☆ Widely reported chromosome no. in okra 2n= 130
- ☆ Total no of cultivated spp.=6
- ☆ Mainly three species are cultivated from (*A. esculantus, A. monihot, A. moschatus*)
- ☆ Okra is facultative autogamy type
- ☆ Photoperiodism =Day neutral
- ☆ Okra is classified as often-cross pollinated crops due to protogyny
- ☆ Natural cross pollinated about 5-12 per cent
- ☆ Summer season crops 18-20 kg/ha
- ☆ Rainy season 10-12 kg/ha.
- ☆ Hybrid seed rate 3.75-5.0 kg/ha (test weight 50gm)
- ☆ Seed viability 2 year
- ☆ seed yield 1200-1500 kg/ha
- ☆ Standard germination 65 per cent
- ☆ Highly pure seed 99 per cent
- ☆ Seed of okra will not germinate bellow 20 ^{0}C
- ☆ Minimum night temp.-22 ^{0}C
- ☆ Okra seed germinate 25-35 per cent but fast germination 35 per cent
- ☆ Temp above 42 ^{0}C cause flower drops
- ☆ Opt. temp for seed germination 29 ^{0}C
- ☆ Storage temp. 8.9 ^{0}C 90 per cent RH, 2 weeks
- ☆ 1st harvesting after 45 days
- ☆ Enhance self-life of okra CCC @100gm
- ☆ Export standard 6-8cm
- ☆ Seed settering is major problem in okra for seed production
- ☆ Hermaphrodite flower, superior ovary, Okra caducous calyx
- ☆ Soil acidity less tolerant (6.8-6.0)

- Respiratory rate very high, quantity of natural removal moderate
- Isolation distance 400 Foundation (white 15 x 7.5, 99.5 per cent pure), 200 Certified (azure blue, 15x 7.5, 99.0 pure)
- BT okra developed by MYHCO company
- Flower bud appear in the axil of each leaf above 6- 8 leaf
- 4-6 day after pollination great increase of fruit weight length diameter
- Fibre formation start 5-6th day after fruit set
- B.T okra resist. To shoot and fruit borer gene **(*Cry-1 AC*)**
- Yellow vain mosaic virus controlled by (**2 complementary gene)**
- Gimini virus reported by C.S kulkarni 1924 (**YV1 and YV2)**
- Ratoon cropping can be donc in okra
- GMS found in okra
- Var. suitable for ratooning Arka Abhay, Pusa A-4
- Pusa Sawani is adopted to large pH has tolerance to salinity
- Field resist. line IC- 1542
- YVMV of okra interspecific cross (*A. esculantus* x *A.monihot*)
- Emasculation done in late afternoon,Dehiscence 8-10am
- Hybrid seed production (female male rartio) 4: 1
- One male flower can used pollinated 3-4 female flower
- Purline var.-
- Pusa Makhmali, CO-1 (selection from red wonder)
- Pedigree method- Pusa Sawani, Punjab Padmini, Prabhnai Kranti, P-7, Arka Anamika, Akra Abhay
- Hybrid seed of okra commercial scale Rannebenur (Karnataka), Maharashtra (Buldana dist.)
- Pusa Makhmali developed by H.B Singh
- Pusa Sawani (IC 1542X P. makhmali)
- CO-1 selection red wonder
- MDU-1 Selection Pusa Sawani (mutant var.)
- Punjab Padmini (*A. esculantus* x *A.monihot* spp. manlhot) (rashmi x ghana)
- Harbhajan/perkins long green or selection-6 introduced from USA (it is notified by central variety release committee)
- Prabhani Kranti (Pusa Sawani x *A. monihot*)
- P-7 (PUNJAB -7)- {*A. esculantus* (Pusa Sawani) x *A.monihot* (ghnna)}

- Selection 10 (Arka Anamika) (*A. esculantus* x *A. tetraphylus*)
- EMS-8 (Punjab-8) mutant of Pusa Sawani
- Versha Uphar (HRB-9-2) (Lem sel. x Prabhani Kranti) intra varietal
- Hissar Unnat (HRB-55) (sel. -2-2 x Prabhani Kranti) intra varietal
- Pusa A-4 Substitute of Pusa Sawani (interspecific var.) (*A. esculantus* x *A. monihot*)
- Kashi Pragati (VRO-6), Kashi Mohini (VRO-3),
- Punchali Adhunik, Supriya, Varsha
- Suitable for export (Pusa Sawani, Pusa A-4, Prabhani Kranti, Versha Uphar)
- Yield loss by YVMV 50-90 per cent
- Jassid is most serious pest
- *A. monihot; A. tetraphylus*- Tolerent to YVMV
- Enation leaf curl 1st reported at (IIHR-1984)
- Progenitor of cultivated okra (*A. tuberculatus*) (2n=58)
- Putative ancestor of okra (*A. tuberculatus*)
- West African okra guinean okra (*A. Callei*) var.- Pusa A-4, multi resistant
- Ghanna okra (complex poly sp.) (*A. Manihot*)
- Highest Ch. No. of okra *A. monihot* spp. Manihot 2n =194
- Symptomless resist to YVMV spp. (*A. tuberculatus*)
- Immune to YVMV (*A. monihot* spp. Manihot)
- True resist to YVMV (*A. pungen*)
- Aromatic spp. (*A. moschatus*) (ambrate)
- Resistant to low temp. (*A. angulosus*)

Chapter 11

Drumstics/Moringa/Horse Radish Tree/Radish Tree/West Indian Bean *Moringa olefera*; Family: Moringaceae

- 2n=2x=28
- Origin-North-West India
- Edible part -Pod
- Tropical plant/deciduous plant (8-10m height)
- Monogenic family
- **Seed oil content is also called as- behen/ben oil**
- Percent of seed oil 18.9 to 38- 40 per cent
- Highly cross pollinated crops heteromorphism, honey bee carpenter bee.
- Flower colour white and creamy white
- Diocious nature
- Propagation: stem and limb cutting (1-1.5m) thickness 14-16cm
- Annual type seed 625gm/ha
- Spacing for annual 3 to 4.5m pant to plant and row to row
- Exporting variety of moringa- villaypatti

- Fruit contain 10-15 seeds
- Hairy caterpillar cause defoliation control- fish oil
- Anthesis time 4.30 to 6.30am (5.30 peak)
- Temperature 27.3 ^{0}C to 29.3 ^{0}C, 68- 78 per cent RH
- Anther dehiscence 4-6: 30am
- Konkan ruchira selection form vasai local best quality
- PKM-1 and 2 annual type
- Chemmurangai, murangai, Kattumurangai, Punamurangai, Moolanoor-murangai, Chevakcherimurangai, Kodaikalmurangai, Palmedumurangai
- PKM-2 (MP-31 X MP-28)
- Pollen viability is 72 per cent at anther dehiscence
- Jaffna/yazloanam murangi - It is suitable south Indian and coastal region (400 pod/year/plants)
- Sajna one time flowering (Jan-Feb), najna tree time flowering.
- Antimalarial properties Tocoferol (7.5mg)
- Rubefaciene properties bitterness 80-90 per cent oil used for cosmetic root used for skin disease
- Better growth developed 25-35 ^{0}C
- Highly susceptible frost
- Seeds are rich in himine/β-carotene 11300 IU/α-tocoferol 43.6gm/kg protein
- Iodine rich source of murangi- 51μg/gm nicotine acid 0.8mg/gm ascorbic acid 220mg/gm
- February-March 1st flowering; and September- November 2nd flowering
- High temp. exceeding 40^{0}Ccause flower shedding harvesting annual 6 month after sowing
- Perennial 8-9 month each tree 200-250 fruit/annual
- 80-90 fruit/perennial

Chapter 12

One Liner Question

1. Cucumber is rich source of = **Chlorine**
2. The direct effect of pollen on the carpel tissue= **Carpoxenia**
3. White lemon is a variety of **=Cucumber**
4. Pickle preserved by oil**= Cauliflower**
5. Chromosome number of cowpea=2n-22
6. Asparagus is a **Cross pollinated crop**
7. Pinching of male flower followed in =**Pumpkin**
8. Sreee Nandini is the variety of which crop-**Sweet potato**
9. Gaur Gum found in **cluster Bean**
10. Bidhan Kusum is the variety of **Elephant foot yam crop**
11. Arka series of variety released from=**IIHR**
12. Seed rate of cucumber - **2.5 kg**
13. Seed rate for pumpkin - **4.5 kg**
14. Edible part of cabbage**- Head**
15. Edible part of Brussels sprout-**Enlarged axillary bud**
16. In which country TPS method is mostly used - **China**
17. Seed plot technique is developed by - **Puskarnath**
18. Arka Nishant is the variety of **Radish crop**
19. Tetrazolium test is done for **Seed viability**

20. 90 per cent microorganisms are killed= **D value**
21. Seed colour of ridge gourd= **Black**
22. Scurvy due to deficiency of-**Vit.-C**
23. Tomato rich in vitamin **Vit-C and Vit- A**
24. Parsley rich in vitamin-**A**
25. Inflorescence of potato- **Cymose**
26. Palampur Green is the variety of _**Chinese Cabbage**
27. Major component found in garlic- **Diallyl disulphide**
28. In radish, pink colour is due to presence of- **Anthocynin**
29. Family of carrot-**Apiaceae**
30. Family of Beet root- **Chenopodiaceae**
31. Little leaf of brinjal is due to- **Mycoplasma**
32. Year for seed quality control-**1983**
33. Botanical name of orange melon- ***Cucumis melo var.hime***
34. Which method is used in cabbage for production of foundation and **certified seed**
35. Isolation distance in tomato for breeder seed-**50m**
36. Kuroda is variety of - **Carrot**
37. Yield of Brussels sprout -**10-16t/ha**
38. Parsley seed rate - **250-300g/ha**
39. PKM-1 released from- **Periyakuluman**
40. Heterosis term coined by- **Shull**
41. Whiptail of cauliflower due to deficiency of- **Mo**
42. Toxic substance present in brinjal **Solasodine**
43. Fruit type of cucumber- **Pepo**
44. In cowpea, isolation distance for certified seed- **25gm**
45. India's share in vegetable production- **14.1 per cent**
46. In tomato Breeding work done by **Dr. C.M Rick**
47. Shelling per cent in garden pea **35-50 per cent**
48. In arid region of Rajasthan which vegetable is produced-**Lasoda**
49. Vegetables are preserved by **Brined** solution
50. Edible part of cauliflower- **Curd**
51. Waxing is done in ________ (a) okra (b) bitter gourd**(c)brinjal** (d) all of above

52. For seed production musk melon is not cultivated with ________
53. Botanical name of field pea- ***Pisum sativum var.arvens***
54. Storage temperature for Globe artichoke- **0°C and R.H 90-95 per cent**
55. Recalcitrant seed found in **Tomato**
56. Alliaceae and amaryllidaceae linked to **Same family**
57. Family of amaranthus- **Amaranthaceae**
58. Edible part of knolkhol- **Brassicae** swollen bulb like stem
59. Family of Rhubarb - **Polygonaceae**
60. Salinity tolerant crop- **Tomato, turnip, bittergourd**
61. Example of climacteric vegetable- **Tomato, brinjal, muskmelon**
62. Toxic substance found in peas and beans - **Phytic acid and trypsin inhibitor**
63. Vitamin A rich variety of pumpkin - **Arka Chyan**
64. Pollination type of brinjal **Day neutral**
65. Progenitor of onion- ***Allium vavilovii***
66. Example of Cleistogamy vegetable- **Lettuce, pea and bean**
67. Highest inbreeding depression found in - **Carrot**
68. Protoandry found in- **Onion, carrot, muskmelon, pointed gourd**
69. Example of multi seeded berry- **Chilli**
70. CH-1 var. of chilli developed by using - **GMS**
71. ***Trichosanthes dioca*** is scientific name of- **Pointed gourd**
72. Andromonoecious and monoecious sex form found in -**Water melon**
73. Iron rich vegetable - **Amarenthus**
74. Potato rich in - **Starch**
75. Storage temperature for tomato- **7.2-10°C R.H 85-96 per cent**
76. Kashi Amrit is the variety of-**Tomato**
77. Kashi Aman is the variety of-**Tomato**
78. Important viral disease in tomato-**TLCV**
79. Tomato fruit cracking is due to- **Boron**
80. Sex form of *Momor dicadioca*-**Monoecious**
81. Kashi petha is the variety of **Ash Gourd**
82. Chemical found in pointed gourd- **Salicylic Acid**
83. Pungency in chilli is due to -**Capsacin**

84. Temperature for production of elephant foot yam=**-25-35°C**
85. Definition of self incompatibility- **Inability of pollen grain of same flowers to fertilised the same stigma of a bisexual flowers.**
86. Arka Harit is a improved variety of – **Bitter gourd**
87. Broad bean pollinated by-**Insect**
88. Pollinator of cucumber **-Cucumber**
89. PusaUdit is the variety of- **Broad bean**
90. Martha Washington is the variety of- **Asparagus**
91. Arka Bahar is a improved variety of **– Bottle gourd**
92. Protein rich vegetable **Pea, Cowpea**
93. Example of perennial vegetable- **Asparagus**
94. Tejaswini variety of chilli released from-**Private sector**
95. IIVR located at **Varanasi**
96. Seed rate of leek - **5-6 kg**
97. Formation of Abscission layer is due to **Auxin** growth regulator
98. Kashi Ganga is an improved variety of **– Bottle gourd**
99. Adaption of new variety in new place- **Introduction**
100. Parthenocarpy in cucumber is due to- ***Pat* gene**
101. Maturity indices for leek- **Neck bent**
102. Kashi Nandini is the variety of- **Pea**
103. Terminator is variety of -**Cucumber**
104. Botanical name of parsnip ***- Pastinaca sativa***
105. Stem cracking of celery due to - **Boron Deficiency**
106. Blindness of cauliflower is due to **poor fertility of the soil**
107. Most devasting disease of okra - **YVMV**
108. Seed viabily of turnip- **5-6 year**
109. Inflorescence of coriander- **Umbel**
110. Vegetable recommended for diabetic patient- W**hite brinjal**
111. Pointed gourd is propagated by - **Stem cutting**
112. Kaempferol is found in **- French bean**
113. Genome size of carrot**- 480 mbp**
114. Country leading in transgenic plants-**USA**
115. Lycopene content in tomato**- 20-50mg/100gm**

116. Botanical name of tree tomato- ***Solenum betaceum***

117. Pusa udit is a variety of - B**road bean**

118. Ripening mutants of tomato- **Polygalacturonase**

119. Onion curing temperature and RH. **0 ^{0}C and 65-70 RH**

120. Seed yield of okra-**1000 kg**

121. Parthenocarpy induced by PGR -**Auxin**

122. Parthenocarpy coined by-**NULL**

123. Antinutrient in brinjal-**Solasodine**

124. Auxin synthesized in site- **Tip of the stem or shoots**

125. Heat unit used to measure maturity of- **Pod**

126. Economic part of celery (Petiole, leaf,none)

127. Pusa Meghdoot parentage **(PSPL x Sel- 2)**

128. Pusa Anmol parentage

129. Pusa snowboll-1 parentage

130. ZECC invented by-**Dr.S.K Roy**

131. Reduction in style length in evolution of tomato suggested by - **Miller**

132. Jade cross is variety of **Brussels sprout**

133. King of north is a variety of- **Knoll khol**

134. Basic chromosome no. of potato -**12**

135. Basic chromosome no. of bottle gourd- **11**

136. Palak seed production takes place in – Summer seasion

137. Potato leaf marker present in which vegetable- **Tomato**

138. Brown seed marker present in-**chinese cabbage**

139. White flowering gourd-**Bottle gourd**

140. Dark yellow corolla present in which gourd- **Bitter gourd**

141. Poisonous sp.of yam-***D.alata, D.esculenta, D.rotundifolia- D.sinnata***

142. 1st order preservative- Sugar, KMS, **Kmno4**

143. Anticancerous compound present in broccoli- ***Indol-3 carbinal***

144. ETL level of potato aphid- **20 aphid**

145. **TPS** is given by-**Dr. S.Ramanujan**

146. Marrow and squash is common name of-**Summer squash**

147. Size of brusselsprout require for freezing-**3cm**

148. Isolation distance for Foundation seeds production of cabbage-**1600M**

149. Calculate the no. of progeny, when 8 parents involved in partial diallel cross- 32
150. Growth is measured by which instrument-**Auxanometer**
151. Colour of Breeder seed – **Golden yellow**
152. Colour of foundation seed-**White**
153. Seed law came into force in the year-**1966**
154. Scarification is done in –**Okra**, Chilli, Cucumber, Radish
155. ***Yerrusuri***and ***fufu*** are processed product of- **Pumpkin and Tapioca**
156. 1ppm= **1 mg/l**
157. 1st developed molecular marker-**RFLP**
158. TPS seed rate-**100-120gm/ha.**
159. Highest area in potato **U.P state**
160. India's rank in area and production of vegetable-**2nd and 14 per cent**
161. Herkogamy found in-**Lima bean**
162. Purple colour flowering chilli spp.- ***C. pubescence***
163. Yellow anther sp. Of Chill- ***C. baccatum***
164. Light reaction occurs in- **stroma and granna**
165. Treatment of vegetables for 2-3 min in boiling water (**blanching**)
166. Selfing and massing in onion proposed by- **HA Jhons and Mann**
167. Overdominance theory given by- **Edward M. East 1908 Jeorgeshull 1908**
168. Colour of viable seed in Tetrazolium test- **Red**
169. Which is an antigibberilin- **CCC**
170. Vernalisation temp. of cole crops **5 to 10 ^{o}C**
171. Okra seed will not germinate below temp.-**5 ^{o}C**
172. Viviparous cucurbit-**chow-chow**
173. Triploid sp. of potato **(*S. chaucha*)**
174. Sterile diploid-**Garlic**
175. Dioecious perennial- P**ointed gourd**
176. Sex form of cassava - **Monoecious**
177. Asparagus is -**Dioecious perennial**
178. Gajendra is a variety of- **Elephant foot yam**
179. 21st century vegetable-**Chekurmanis**

180. Low temp. legume- **Broad bean**

181. Paraheliotropism found in**-** ***Phaseolus vulgaris L.***

182. BER of tomato is due to –**Ca Deficiency**

183. Petha prepared from-**Ash gourd**

184. Kanji prepared from **Black** colour carrot

185. Disorder of potato due to high spacinghigh N_2- **Hollow heart**

186. Pattypan released from- **USA**

187. Which colour of brinjal preferred for diabetic people- **White**

188. Which colour of onion used for dehydration- **White**

189. Round, light green colour seed of french bean preferred for-**String beans or snap bean**

190. Quick seed germination test is coined by-**Tetrazolium test George Lakon**

191. Seed rate of knol-khol-**1.5kg**

192. Phytochrome is related to flowering, germination- **Control many aspect of plants development**

193. Triploid seed production is given by-**Dr. Kihara in watermelon**

194. Anti—Sterility vitamin**- Vit-E**

195. Deficiency of protein causes disease-**kwashiokor and marasmus**

196. Inflorescence Of cabbage-**Catkin**

197. Yellow colour of onion is due to**-Quercetin**

198. Anti-fungal compound present in onion-**Phenolic compound**

199. Chinse yellow is a variety of- **Cucumber**

200. Tight head is present in which types of lettuce- Ice berg **(Crisphed lettuce)**

201. Most suitable soil for vegetable production**- Loamy soil**

202. Self blanching term is related to-(a) Cauliflower; (b) Lettuce; (c) Cabbage; **(d) Both a and b**

203. Seed extraction method of tomato- Alkali, Acid, Fermentation**, All**

204. Inflorescence of amaranthus- **Compound spike**

205. Vegetables belongto which division of flowering plants-**Angiosperms**

206. Pusa lal meeruti is developed through which ray mutation- **X- ray**

207. S.N of parsnip- ***Pastinaca sativa***

208. Family of celery-**Apiaceae**

209. Family of parsley **-Apiaceae**

210. Central food laboratory is situated at- **Kolkata**

211. TSS of tomato paste and purry- **Puree TSS 9-12 per cent and Paste TSS 25 per cent**

212. Tomato is introduced to India on which year **16th century**

213. Plastic mulches are used for-controlling weeds, control soil temperature, control moisture, **all**

214. Thickness of plastic cover used for soil sterilization- **100 micron**

215. AICVIP was started on which year-**1970-71**

216. In cucurbits seed production programme which of the lines are used- GYN,TRI,HERM,**ALL**

217. The autocatalytic function of ethylene is associated with – **Growth and senescence**

218. Maturity indices of Rabi onion - **Neck Fall**

219. Origin of Ceylon spinach – **Asia (India to new guinea)**

220. Central Asia is primary centre of origin of which crops - **Onion**

221. Seed multiplication ratio in radish – **1: 100**

222. Genomic analysis through RFLP and SSR helps to – **Polymorphism**

223. PPV and FR bill was passed in-**2001**

224. PPV and FR Act 2001came into force in-**2005**

225. For breeder seed, which method of seed production in carrot and beet

226. Isolation distance for certified seed in cabbage – **1000 m**

227. Isolation distance for certified seed in beetroot – **1000 m**

228. Wind pollination in – **Leafy vegetable**

229. Row to row spacing of Basella – **60 x 20-30 cm**

230. VAM provide which mineral to plant- **P**

231. N_2 fixing microbes in legume – ***Rhizobium spp.***

232. French bean is – **Green bean**

233. Definition of photoperiodism- **Response of light on plants for flowering behaviour**

234. Seed rate of leek - **5-6 kg/h.**

235. Basic chromosome number of rhubarb – **n=11**

236. Chromosome number of bitter gourd – **2n = 2x =22**

237. Chromosome number of Sponge gourd – **2n =2x = 26**

238. Seed colour of sponge gourd - **White**

239. Genetics of andromonoecious in muskmelon- **aa**

240. Conditions for flowering in potato- **SDP**

241. Conditions for tuberization in potato -**20 °C**

242. Tuberization in sweet potato **Temp- 20 – 30 °C Light intensity – 1800 – 40000 Lux.**

243. Storage temp of potato for processing – **2 – 4 °C and RH -90-95** per cent

244. Fruit set temp in tomato – **24-27°C**

245. RFLP is molecular marker - **Hybridization based and co- dominant markers & first molecular marker**

246. 1st purple colour variety of radish **Pusa Sagarika**

247. RAPD is a PCR based - **Dominant marker**

248. Variety of carrot suitable for canning - **Chantenay**

249. Late blight is maximum at - **66°C and RH-90°C**

250. Phytopthoraporri cause which disease – **Root rot disease**

251. Onion temp for bolting – **20 – 25 °C**

252. Bolting in lettuce – **Plant has matured and**

253. Early planting of kharif onion and late planting of rabi onion resulted in bolting

254. Leaves of brinjal contain more solodosinethan fruit

255. CGMS in pepper is commony used – **ms-10**

256. GMS and CGMS commercially exploited in - **Chilli**

257. Seed yield of brinjal – **150 – 200 kg/h**

258. Seed yield of okra – **1000 kg/h.**

259. Cold temp requirement for Indian cauliflower – **15 -20 °C**

260. Powdery mildew in garden pea controlled by - **Fungicides**

261. Puffiness in tomato due to – **Incomplete pollination**

262. BER in tomato – **Calcium deficiency**

263. BER in water melon- **Calcium**

264. Moncot family *e.g.* – **Amaryllidaceae, Araceae, Graminae, Liliaceae**

265. Example of monocot vegetable **Onion, Garlic, EYM, Tania, Sweet corn, Aspragus.**

266. Temperature for midlate cauliflower – **12 – 16 °C**

267. Fruit of okra – **Capsule**

268. Family of Chinese potato – **Lamiaceae**

269. Queensland arrow root – **1 starchy edible rhizome**

270. Naturally occurin cytokinin – ***Arabidopsis***

271. Salt act as preservative through – **It reduces the water activity of food.**

272. PCR techniques was given by – **Kary Mullis (1985)**

273. Ethylene absorbant–**$KMnO_4$**

274. Ethylene harmfulness based on – **Skin contact headache dizziness fatigue**

275. Increase of root due to – conc of IAA - **IAA higher concentration inhibits root growth.**

276. Blotchy ripening in tomato is – **Boron deficiency**

277. Pungency in chilli – **Capsacin**

278. Curing temp of onion – **20 ^{0}C**

279. Removal of water by applying heat – **Evaporation**

280. Polyploidy var. of palak – **Pusa Jyoti**

281. Mutant var. of French bean – **Pusa Parvati**

282. YVMV transmitted by - **Whitefly**

283. Tospo virus transmitted by **– Thrips**

284. Definition of apomixes **–Development of seed without true fertilization.**

285. Edible part of Jerusalem Artichoke **– Tuber or the swollen end of tha underground stem.**

286. Pole type var. of lima bean- **Florida Butter, Chalenge, and King of tha Garden**

287. Fufu is processing product from - **Casava**

288. Highest inbreeding depression – **Lucerne, Carrot**

289. Selfing and massing is done in – **Bulb Crops.**

290. Edible part of Knolkhol – **Knob or Tuber**

291. Lime loving crop – **Legume such as broad bean pea, onion garlic asparagus**

292. Scab in potato prevails – **Caused by bacteria like organism steno-phagous scabies**

293. Paraheliotrophycondition is found in which crops- **Cowpea**

294. Reason for hollow heart in potato – **Excess nitrogen**

295. Cross of triploid water melon – **Pusa Bedana (Tetra-2 x Pusa Rasal)**

296. RAPD is a **Dominant markers.**

297. Multivitamin green-**Chekurmanis**

298. Shallow rooted vegetable – **Onion, Garlic, Lettuce, Spinach, Cabbage, Broccoli**

299. Crop whose seed germination slow-**Lima bean, Parsley, Watermelon, Okra**

300. Example of protandry-**Carrot, onion, beet, parsley, leek**

301. Moringa also known as- **Drumstick**

302. Rotting in vegetable-**Air, moisture, light, temperature, microorganism**

303. Onion dehydration ratio-**10: 1**

304. Little leaf of brinjal transmitted by-**Leaf hopper**

305. Flower of brinjal-**Pentamerous**

306. Basella is________crop-**Edible perennial vine and widely used as leafy vegetable.**

307. Potato leaf marker-**Marker assisted selection (MAS)**

308. Tomato subjected to pinching for-**Upright type of tomato**

309. Type of cracking in ripe fruits-**Two types (1) radical (2) concentric**

310. Which stage harvested tomato stored in ZECC for 4-5 days-**Middle ripe stage**

311. Flower cluster in tomato-**Truss**

312. Origin of water melon-**Tropical Africa**

313. Newsland spinach-**Minor leafy vegetables, it succeed in hot weather**

314. Anthesis in single umbel of carrot completed in -**7 to 9 days**

315. Maturity indices helps in-**Harvesting**

316. Fruit yield in brinjal related to-**Variety 30-50 t/hac. Hybrid 50-90 t/ hac.**

317. In bottle gourd high heterosis for-**Increase F1 in fitness and vigour over the parental value.**

318. Storage life of potato at 3.4-4° cand 55 per cent RH **-2-4°C and RH 90-95 per cent**

319. Micronutrient essential for auxin synthesis-**Iodine**

320. Staking-**Tomato pea and beans peppers *etc.***

321. Isogenic lines derived through-**Generated by repeated backcrossing**

322. Very perishable vegetable-**Leafy vegetable**

323. Breeding objective of radish- **Good quality and free from disease and pest**

324. Vibrator used in polyhouse for-**For pollination**

325. Pentaploid sp. of potato – **2n =5x=60 Solanum cartilobum**

326. Ascorbic acid prevent- **Scurvy diseas**

327. ZECC principle-**Working on the principle of evaporating cooling**

328. SPT developed by- **Pushkarnath 1967**

329. Spray of— break dormancy in potato-**1 per cent thiourea + 1 ppm GA3 for 1 hr.**

330. In onion irrigation stopped-**10-15 days before harvest**

331. Antiauxin Known as – **Para chloro phenoxy isobuteric acid (PCIB)**

332. Temp more important than day length— in onion-**11-12 hrs.**

333. Spraying of —— for fruit setting in tomato during summer months-**4-CPA-4 Chloro phenoxy acetic acid (40 ppm)**

334. Nursery area of tomato-**225-250 metre sq.**

335. Percentage of Hissar Anmol-**Hisar Arun X *L. hirsutum f.glabratum***

336. Crop classified as self and cross pollination based on-**Pollination**

337. Cherry tomato exported in-**Turkey, Italy, India**

338. Most of leafy vegetables and root crops rich in-**leafy veg.-nutrition, root crops –caretenoids**

339. Definition of male sterility- **Non functional of male organs of a bisexual flower**

340. Inbreed line having same genetic constitution but differ at one locus-**Multiline**

341. For improving head size and vit C in cabbage which breeding method used- **Family selection**

342. Pie plant-**Rhubarb**

343. French bean having light green colour-**Greenbeans**

344. Perennial leafless plant-**Horse tail**

345. Coriander family-**Apiaceae**

346. Chemical compound present in cluster bean- **Gallotanis, Galic acid**

347. Maturity indices of musk melon-**Netting**

348. Sugar heated above melting point-**Sucrose crystals lies between 11-19 °C melting point**

349. Vegetable having maximum share in export-**14 per cent in world**

350. Mineral elements present in leafy vegetables –**Fe and Ca, vitamins (A, C, K and riboflavin)**

351. Minimum germination per cent of cauliflower-**65 per cent minimum**

352. **Fermentation** is a permanent method of preservation.

353. In potato hulm cutting is done to reduce aphid population and drying of periderm skin

354. After exhausting in canning for heating products passed through water bath

355. Blanching done in **Celery**

356. Spinach and Broad bean are having the same chromosome number **2n=12**

357. Short duration Radish Var. **Pusa Rashmi**

358. Most sensitive vegetable to ethylene injury is **Cucumber**

359. Ploidy level of Okra is Aneuploidy, Polyploidy and **Tetraploidy**

360. **Seed Plot Technique** method of seed production takes place in North India Plains.

361. **A. Ujjaval** is the variety of multiplier Onion.

362. Parsley is **protendry**

363. **2, 4-D** is commonly used as weedicide.

364. Pinching is commonly followed in **Cucumber**.

365. Temperory method of preservation: **freezing**

366. **Globe artichoke** propagated by suckers, crown divisions and off shoots

367. **Sweetpotato** is hexaploid.

368. Puffy fruit is a physiological disorder.

369. Tender fruit (8-9cm length) of Okra is best suitable for export.

370. At the time of leaf falling auxins increases.

371. Di-phenyl amine found in onion is effective against diabetes

372. Vegetable consumption of an Indian per day 145g is far less than recommended 300g.

373. Per capita consumption of vegetables in India is 62-79kg/annum as compare to Greece 377, Turkey 327, Republic of Korea 216, China 215 and Japan 105kg/annum.

374. AICRP on vegetable crops started in 1971-1972 with 17 centers (7 main centers and 10 sub-centers) at IARI New Delhi.

375. Varieties are released for cultivation by CVRC (Central variety release committee).

376. The first F1 hybrid released from IARI was Pusa Meghdoot in Bottle gourd in 1971.

377. Indo American hybrid seed that's released its first tomato hybrid 'Karnataka' and first capsicum hybrid 'Bharath' in 1973.

378. Two synthetic varieties in cauliflower are developed Pusa synthetic in 1975 and Pusa early synthetic in 1990.

379. Tomato variety for cold set 'Pusa sheetal' hot sett 'Pusa hybrid 1' and for both (cold and hot) is 'Pusa Sadabahar'.

380. Pusa purple long is the variety of Brinjal is popular all over the country.

381. Munjari gota and BG-1 is the popular varieties of Brinjal for 'Bharta' making.

382. Pusa Sawani resistant to YVM virus in okra released in 1957 was first successful example of resistance breeding.

383. Annamalai in brinjal resistant to aphid.

384. Root knot nematode resistant in tomato varieties is: Sel-120 released in 1966.

385. Pusa Jwala resistant to chilli virus released in 1972.

386. Parthenocarpic varieties in tomato are Severianian.

387. In onion use of MH and NAA recommended for suppression of sprouting, rotting and weight loss in cold storage and extending their storage life.

388. Red rot of cauliflower is caused by Boron deficiency.

389. Blossom end rot in tomato due to moisture stress.

390. India known as 'Land of spices' grows around 63 different types of spices and is the largest producer, consumer and exporter of spices in the world.

391. Jammu and Kashmir produces world best saffron and Punjab produces 90 per cent of celery produced in the world.

392. Seed spices include coriander, cumin, fennel and fenugreek.

393. AICRP on spices was initiated in 1971 but later on in 1986 its headquarter shifted to NRC Spices Calicut.

394. AICRP on Betel Vine was started in 1981.

395. NRC for spices came into existence in 1986.

396. Directorate of Areca Nut and spices development Calicut was established on 1st April 1966 at Calicut.

397. Katte virus disease found in cardamom.

398. Nvashree and Nithyashree released from IISR and two other Cinnamon common varieties are Kokan Tej and Yercud-1.

399. Pepper is the most important spice of India and known as 'King of Spices' and 'Black gold of India'.

400. India is the major producer, consumer and exporter of chilli in the world.

401. India is the largest producer of turmeric (90 per cent) in the world.

402. All vegetables belong to sub-community spermatophyte which includes two division, gymnosperm and angiosperm.

403. Most of the vegetables belong to Dicotyledonous family and few belongs to monocotyledonous like Amarylidaceae, araceae, Dioscoraceae, Liliaceae and Poaceae (Gramineae).

404. The name of cabbage and cardamom inflorescence is catkin.

405. Pollen is transfer from anther to stigma of the same flower is called self pollinated or autogamy.

406. Pollens are transferred to stigma of the other flower or cultivar or spices are called cross pollination or allogamy.

407. Broccoli has the highest cross pollination rate i.e., 95 per cent.

408. Gametophytic self incompatibility found in tomato.

409. Tomato, musk melon, watermelon comes under climacteric while, cucumber comes under non-climacteric fruits.

410. Concept of centre of origin was given by Vavilov.

411. All centres of origins are the centre of diversity, but all centre of diversity are not centres of origin.

412. Pusa Bedana in water melon and Dasheen in Colocasia are found in triploid (2x=3n) in nature.

413. Okra is tetraploid (allotetraploid).

414. Sweet potato and great headed garlic are hexaploid.

415. Amaranthus and globe artichoke are C4 plant.

416. Red colour in root of beetroot is due to β-cyanine and yellow colour is due to β-xanthin.

417. Bathua is also known as pigweed.

418. Satputia is a hermaphrodite variety of ridge gourd.

419. India has the total geographical areas of 328.73 million hectare.

420. Major soils of India are red sandy (15 per cent).

421. Perfection is the variety of Asparagus.

422. Pusa visesh, arka harit and pusa mausami is the variety of bitter gourd.

423. Cabbage variety Pusa synthetic, September, golden acre, pride of India, Copenhagen market.

424. Capsicum/sweet pepper- California Wonder, king of North, World Beater, Arka Mohini.

425. IIVR-PG-105 seedless variety of pointed gourd.

426. Sweet potato- Kalmegh, Pusa Lal, Pusa Safed.

427. Tomato var. Sioux, Pusa Sheetal, Italian Red Pear, Marglobe, Best of All, Fire Ball, La Bonita, Gulmohar, Punjab Chuhara *etc.*

428. Water melon var. Pusa Bedana, Madhu, Santrupi, Sugar Baby, Durgapur Meetha, Durgapure Lal.

429. Bower system of training was most efficient in increasing the productivity of bitter gourd. Amaranthus is cross pollinated C4 plant.

430. Asparagus tender shoots contain a chemical 'asparagines' which is used as a medicine as diuretic.

431. Four types of flower found in brinjal are longe style, medium style, pseudo-short style and short style/true short style.

432. Only long and medium styled flower bears fruit. Another two are rudimentary ovaries, which do not develop into fruits.

433. Three recognised grades in brinjal are: super fancy and super commercial.

434. Broad bean is the only bean, which can withstand sufficiently cold temperature.

435. Black carrots are used for the preparation of kanji, a good appetizer.

436. The orange colour varieties of carrot are rich in carotene, a precursor of vitamin A, and contain appreciable quantities of thiamine and riboflavin.

437. Carrot var. pusa kesar, pusa meghali, pusa yamdagni, coreless, chantaney.

438. In India cauliflower crop was introduce during 1822.

439. The name of cauliflower was derived from th latin words caulis and floris which mean cabbage and flower respectively.

440. Cauliflower is originated in Cypress and around Mediterranean region.

441. Chekurmanis known as 21st century vegetables or multivitamin green due to very high vitamin content.

442. Capsacin content in chili varies from 0.2-0.4 per cent.

443. Trinidad Scorpion, Andhra Jyoti is the hottest cultivar if chili.

444. Chow chow is the richest among gourds in nutritive value.

445. Santragachi and Gajendra are non-acrid high yielding variety of Amorphophallus.

446. French bean/kidney bean var. - premier, tender green lakshmi.

447. Garlic is a multiple or compound bulb consisting of 10-20 bullets or segments called cloves.

448. The uninjured bulb contains a colourless, odourless water soluble amino acid 'allin' on crushing the bulb the enzyme 'allinase' acts on allin to provide allicin which contains mainly the odoriferous diallyl disulphide.

449. Pungency in horse radish is due to allyl iso-thiocynate and butyl thiocynate.

450. Kale/karam sag-Brassica oleracea var. acephala

451. Swollen stem called 'knob' firms the edible part of knoll khol.

452. Knoll khol var. Purple Viena, White Viena, King Of North, King Of Market.

453. Leek var. London flag, American flag, Elephant.

454. Lima bean is also known as double bean.

455. Lima bean var. Challenger, King of Garden, Florida Butter, Baby Potato.

456. Pointed gourd are dioecious (staminate and pistilate flower are born on seprate plants).

457. TPS technique given by Dr. S. Ramanujam and SPT technique given by Dr. Pusakarnath.

458. Radish var. French Breakfast, Pusa Reshmi, White Icicle, Scarlet Globe, Pusa Chetki.

459. Ridge gourd var. Pusa Nasdar, Satputia, Rohini, Surekha, Konkan Harita.

460. Broccoli an Italian word, derived from Latin word 'Brachium' meaning an arm branch.

461. Sweet potato is also known as Spanish potato. Turnip var. Pusa Swarnima, Pusa Chandrima, Snowball, Golden Ball, White Globe.

462. Water melon var. Pusa Bedana, Madhu, Amrut, Sugar Baby, Mohini.

463. Winged bean also known as four-angled bean and goa bean in a tropical legume.

464. Cardamom/queen of spices.

465. Katte disease caused by virus spread through banana aphid Pentalonia nigronervosa.

466. Pepper king of spices is leading national spice.

467. All spices *viz.*, nutmeg, clove, cinnamon and pepper.

468. Green cloves are used for pickling.

469. Cinnamon is oldest known spice of India.

470. YVMV in okra is transmitted by white fly (Bemisia tabaci).

471. Okra YVMV is not a seed born disease.

472. Tomato mosaic virus is easily sap transmissible and is principally transmitted by man through contact during cultivation, seed transmission occurs as external seed contamination.

473. Bacterial wilt in tomato/brinjal (Pseudomonas solanacearum).

474. Physiological disorder in colocasia and elephant foot yam are delayed cooking.

475. Pinking disorder in cauliflower is due to anthocyanine formation.

476. Metsubre a nutritional disorder of taro due to calcium deficiency.

477. Color rot in Amorphophallus is due to Sclerotium rolfsii.

478. Brinjal shoot and fruit borer –Leucinodes orbonalis control by spray Cypermethrin 0.07 per cent.

479. Fruit borer in chili and tomato or gram cater pillar- Helicoverpa armigera control by Indosulphon 0.07 per cent.

480. Red pumpkin beetle- Aulocophora foveicollis.

481. Shoot and fruit borer in okra- Earias vittella.

482. The fruits of chili and pepper are botanically 'berry'.

483. Sweet Banana and Bull Nose are introduce variety of chili.

484. Okra is often a cross pollinated crop, 2n=130, n=52, 56, 72, 90.

485. Allium cepa (2X) X Allium fistulosum

486. Allium cepa var. viviparum Cv. Pran (3X)

487. Asiatic type of cultivar are red coloured because of anthocyanin and Europian types are orange because of carotene a precursor of vitamin A.

488. Carrot is a cross pollinated crop.

489. Broccoli an Italian word derived from Latin 'Brachium' meaning an arm of branch.

490. Pant Rituraj a thermosensitive variety of Cowpea.

491. In 1981 the first F1 hybrid of brinjal released by IIHR-Arka naveen.

492. Highest areas under vegetables are tomato-29 per cent, cabbage-23 per cent, brinjal-14 per cent, okra 5 per cent.

493. Cauliflower is a cross pollinated crop.

494. Male sterility line in carrot is reported by Welsh and Grimbell in 1947.

495. Red onions are tolerant to smudge due to the presence of catechol and proto-catechol.

496. Home gardening is also known as Amateur gardening.

497. Choralin present in bitter gourd and diphenylamine found in onion are effective against diabetes.

498. Rajgira leaves are richest source of vitamin A fallowed by Colocasia leaves.

499. Radish leaves are richest source of riboflavin fallowed by fenugreek leaves.

500. Tapioca is the richest source of carbohydrates and calories fallowed by sweet potato.

501. Cluster in flower in tomato is known as truss.

502. The flaver saver tomato is first genetically engineered. Whole fruit reached to US consumer. It was developed by 'Calgene fresh' a California based biotechnological company.

503. Solanum esculentum variety bears egg/round shaped fruit.

504. Solanum serpentinum have long slender type fruit and Solanum depressum plants are dwarf type and fruits are small, purple and pear shaped.

505. Dry chili contains 6 per cent shell, 40 per cent pericarp and 54 per cent seeds.

506. Capsicum annum and Capsicum frutescence have white flower while, Capsicum pubescence has 'purple flower'.

507. Cauliflower was introduced in India in 1822 by a botanist Dr. Jemson from Kew garden in London.

508. **Khol rabi** is Japanese name of Knol khol.

509. The inflorescence of broccoli is cymose.

510. Rutabaga (Brassica napobrassica) is also known as Swedish turnip.

511. Akashin is a physiological disorder of radish caused due to boron deficiency.

512. Netherland is a leading exporter of onion contribution about 21 per cent of world export.

513. Spraying of MH 2500ppm at 75-90 days after transplanting checks the sprouting in onion.

514. Arkel is the most important and popular variety in India introduced from USA.

515. Pea proteins are deficient in sulphor containing amino acids, methionine and cysteine.

516. An alcohol insoluble solids (AIS) range of 11-16 per cent is ideal for canning of peas.

517. Indian bean/Dolichus bean/Hyacinth bean is a self pollinated crop.

518. Lima bean is also known as double bean or butter bean.

519. Broad bean also known as faba bean or horse bean.

520. The allergy of some people to the pollen and green pods of broad bean is known as 'Favism'.

521. Cucumber is monoecious and male flowers appear first fallowed by female flowers.

522. The first F1 hybrid pusa sanyog developed in India in 1971 has a gynoecious female parent.

523. Bitter principles in cucurbits are cucurbitacins which are tetracyclic tri-terpenes.

524. Metaxenia: bitterness in cucumber is due to metaxenia effect. When bitter pollens fertilize non-bitter ovules, the resulting fruits become bitter. This phenomena is called metaxenia.

525. Musk melon is climacteric fruit.

526. Dr. Khiara, a Japanese scientist is associated with triploid (seedless) water melon production.

527. Bitter gourd is also known as balsam pear or bitter cucumber.

528. About 2000-2500 cuttings are required to plant one hectare of pointed gourd and sweet potato.

529. The flower of ridge gourd open in afternoon while, those of spong gourd opens in morning.

530. Ridge gourd contains gelatinous compound called luffein.

531. Chayote/chow chow/cristophine/choko is single seeded vegetables.

532. In India potato was introduced in 1615 by Portuguese. Research on potato started in 1935 at IARI and the CPRI Kufri.

533. Solanum tuberosum is an allotetraploid.

534. Potato is a dicotyledonous and self pollinated.

535. About 40-45g true potato seed (TPS) is enough to plant one hectare.

536. CPRI (Central potato research institute) was established in 1949 with headquarter at Patna which was later shifted to Shimla in 1956 for potato breeding purpose.

537. International Potato Centre (CIP) is an autonomous scientific Institute at Lima in Peru, established in 1971.

538. The degradation of starch into sugars at low temperature is catalysed by the enzyme phosphorylase whereas the reverse process at high temperature is catalysed by the enzyme starch synthase.

539. White star variety of sweet potato with higher starch contain in principally used for processing.

540. Sree shilpa is the first hybrid variety identified and released in Dioscorea alata.

541. The bitter principle in cassava is Cyanogenic glycoside (HCN).

542. The mature pods of yam bean containing seeds are source of a toxic substance rotenone and are sometimes harmful for grazing cattle's in the field.

543. Pusa chaulai is the major cultivar of Amaranthus.

544. Amaranthus is cross pollinated anemophillous crop.

545. Fenugreek is the third largest seed spice in India after coriander and cumin.

546. Fenugreek seeds substantially contain diosgenin which is used as starting material in synthesis of sex hormones.

547. Perfection is the improved variety of asparagus released at IARI New Delhi.

548. The edible portion of asparagus is spear, which is a stem (cladphylls) arisen forma bud on rhizome. The crown is made of roots and rhizomes, which are underground stem.

549. Yellow heart is a variety of bread fruit (Artocarpus altilis) is native of Malaysia.

550. Suwasini is a variety of curry leaf.

551. Rajasthan produce 80 per cent coriander and 90 per cent of cumin produced in the world.

552. Seed spice includes coriander, cumin, fennel and fenugreek.

553. Vietnam is the largest producer of black pepper and cardamom.

554. Pepper research Institute at Penniyur Kerala established in 1953.

555. Small cardamom (Elletaria cardamomum), large cardamom (Amomum subulatum) is a shade loving plant (Seophyte) and propagated through suckers.

556. Kokan Tej and Yerrcud-1 is the variety of cinnamon.

557. 'Allepy turmeric' quality is the best in the world.

558. India is the largest producer of dry ginger in the world (33 per cent).

559. Iran is the major exporter of cumin/zeera (Cumin cyminum L.) in the world.

560. Welsh and Grimbell in 1947 first reported male sterility in carrot.

561. Type of inflorescence in cabbage is catkin or racemose.

562. All the vegetable crops belong to the plant-division angiosperm.

563. Tannia- Xanthosoma sagittifolium.

564. Herkogamy observed in lima bean.

565. Pusa sawani in okra is the first resistant variety released in vegetable in India.

566. Udaipur-102 is onion variety suitable for dehydration.

567. Capsaicin is chemically $C17H27NO3$

568. Little leaf in brinjal is transmitted by leaf hopper.

569. Mosaic in chili is transmitted by white fly.

570. Vaishali vadhu is the variety of okra.

571. Kufri jyoti X kufri alankar- kufri badshah.

572. Cinkle in potato is caused by virus.

573. London flag is a improved variety of Leek.

574. Kovur/Gajendra is an improved variety of Elephant foot yam.

575. Kovur is a improved variety of Colocasia.

576. Roma, vaishali and Punjab chuhara are suitable variety for processing.

577. Glucose-1-phosphate enzymes convert starch to sugar in potato.

578. Allinase is a physiological disorder of cucumber due to calcium deficiency.

579. Frog eye rot is physiological disorder of chili.

580. Metsubre is a disorder of Colocasia due to Calcium deficiency.

581. Tomato was introduced in India by Portuguese.

582. Luckwill (1949) divided to Lycopersicon into two sub-genera-Eripersicon and Eulycopersicon.

583. Dissepiments organ of the chili fruit is richest in pungency.

584. Arka jyoti-water melon, Pusa jyoti- palak.

585. Planning commission divided to climate into 15agro-climatic zones.

586. Bureau of soil survey national land use planning (BSSLUP): 21 agro-ecological region

587. Green leafy vegetables are a good source of niacin and vitamin K.

588. Pusa harit- palak, arka harit- bitter gourd.

589. Sylvia and mithi phalli are the variety of pea which pods are edible.

590. Horse radish tree is common name of drumstick.

591. Coleus is commonly known as Indian potato.

592. Jades cross is a hybrid of Brussels sprout.

593. Chieftain is the variety of Savoy cabbage.

594. Cucumber green mottle virus (CGMV) is transmitted to water.

595. Balsam pear is common name of bitter gourd.

596. Yerrusserri product is prepared from pumpkin.

597. Money maker is the variety of tomato.

598. Winged bean (four angled bean) is the 20th century of vegetables.

599. Inbreed line X Open pollinated variety- Top cross

600. Fufu is an important product made from yam.

601. All India coordinated Potato improvement project started in 1971.

602. Most of the spice belongs to family Myrtaceae.

603. CMS line in pepper (Capsicum annum) was first reported by Peterson.

604. Ramsey and Golsey are important variety of large cardamom.

605. Chirkey and Foorkey is the major viral disease of large cardamom.

606. Amomum sublatum- Nepal cardamom.

607. Black pepper- Pepper nigram

608. Major all spice producing country in the world is Jamaica.

609. Most expensive crop in the world is Saffron.

610. Small cardamom- Queen of spices.

611. 90 per cent pungency present in chili on placenta.

612. Cucumber is indigenous cucurbits.

Chapter 13

MCQ in Vegetable Science

1. **Most suitable time for sowing of Pusa Chetki Variety of radish is:**
   a) March-August
   b) October-November
   c) After November
   d) None of the above
2. **Pusa Chikni is a variety of:**
   a) Ridge gourd
   b) Bottle gourd
   c) Sponge gourd
   d) Bitter gourd
3. **Which of the following variety of Brassica pekinensis is a useful leafy vegetable for plains and hills of North India?**
   a) Chinese
   b) American
   c) Canadian
   d) Japanese
4. **Which of the following type of lima bean variety needs support?**
   a) Pole type
   b) Bush type
   c) Semi-pole type
   d) All of these
5. **Which of the following is a single seeded cucurbit?**
   a) Chayote
   b) kakrol
   c) Kartoli
   d) Ivy gourd
6. **Which of the following is a perennial cucurbitaceous crop?**
   a) Parval (Pointed gourd)
   b) Snake gourd
   c) chow-chow
   d) None of these

7. Protein content of Guar meal is about

a) 11.3 per cent **b) 33.3 per cent**

c) 22.2 per cent d) 44.4 per cent

8. Which of the following pea variety gives 50 per cent produce during first picking?

a) Arkel **b) Mattar Ageta-6**

c) Meteor d) None of these

9. French bean is a native of

a) Asia b) Mediterranean region

c) South and Central America d) None of these

10. Anthracnose resistant variety of French bean is

a) Contender b) Kentucky wonder

c) Tweed wonder d) None of these

11. The best suitable storage condition for fresh unshelled peas is

a) 5°C and 85 per cent RH **b) 0°C and 90-95 per cent RH**

c) 15°C and 60-70 per cent RH d) None of these

12. Okra pods axe rendered unmarketable due to the attack of

a) Aphids b) Mites

c) Painted bug d) None of these

13. Which of the following shape the brinjal flower attain at the time of fruit selling?

a) Short and medium style b) Short style

c) Long and medium style d) Medium style

14. Fruits of vegetable marrow are ready for first picking after

a) 40-50 days b) 80-100 days

c) 10-30 days **d) 60-80 days**

15. Which of the following is the cause for bitterness in wild forms of bottle gourd

a) Solanin **b) Cucurbitacin**

c) Tomatine d) None of these

16. The seed rate of round gourd for one hectare is

a) 3-5 kg b) 8-12 kg

c) 5-8 kg **d) 13-15 kg**

17. Which of the following pointed gourd numbers are enough to plant an area of one hectare?

a) 20,000 b) 30,000

c) 40,000 **d) 50,000**

18. Which of the following type of vegetable gardening is followed in Dal Lake of Kashmir Valley ?

a) Floating garden b) Kitchen garden

c) Hydroponics d) None of these

19. Types of vegetable gardening according to the purpose for which they have been developed are

a) Two b) Four

c) Six d) Eight

20. Durgapura Kesar' is an improved cultivar of

a) Musk melon **b) Water melon**

c) Cucumber d) Bottle gourd

21. Which of the following variety of bottle gourd is without 'crook neck'?

a) Pusa summer prolific long b) Pusa summer prolific round

c) Pusa Meghdoot **d) Arka Bahar**

22. Which of the following is YVMV resistant?

a) Pusa Makhmali **b) Pusa Sawani**

c) Punjab No. 8 d) None of these

23. Which of the following variety of sponge gourd has been recommended from IARI?

a) Pusa Nasdar **b) Pusa Chikni**

c) Yard Long d) None of these

24. Which of the following variety of winter squash has been recommended from IIHR?

a) Arka Chandan b) Large red

c) Arka Suryamukhi d) None of these

25. Which of the following variety is tolerant to fruit borer?

a) Punjab Neelam b) Jamuni Gola

c) Punjab Barsati d) None of these

26. Varsha Upahar is a variety of

a) Cabbage
b) Okra
c) Chilli
d) Cowpea

27. Vegetables are packed in cans and processed by heat in:

a) Retort
b) Sun
c) Retort
d) Any other means

28. Which of the following planting ratio of female to male gives good fruit yielding of parval?

a) 15-20 to 1
b) 30-40 to 1
c) 1 to 1
d) None of these

29. Which of the following planting space of onion bulbs gives high seed yield?

a) 60 x 45cm
b) 45 x 30cm
c) 45 x 45cm
d) None of these

30. Which of the following planting space is recommended for higher yield of onion bulbs?

a) 30 x 45 cm
b) 30 x 7.5cm
c) 15 x 7.5cm
d) None of these

31. Which of the concentration of boron is required to spray the plants at 2-4 leaf stage for higher number of female flowers in bottle gourd?

a) 20 ppm
b) 3 ppm
c) 10 ppm
d) None of these

32. For growing carrot, radish and turnip for table purpose, sowing is done at a spacing of:

a) 45 x 30cm
b) 45 x 7.5cm
c) 30 x 15cm
d) None of these

33. For induction of greater number of female flowers in water melon, spray seedlings at 2-4leaf stage with 2, 4, 5-tri iodobenzoic acid at a concentration of

a) 10-15 ppm
b) 25-50 ppm
c) 50-100 ppm
d) None of these

34. For keeping the sugar content low, potato tubers should be stored at a temperature of

a) 5°C
b) 10°C
c) 15°C
d) None of these

35. For long storage, potato tubers should be kept at

a) 10-15°C b) 5-10°C

c) 15-20°C d) None of these

36. Cucumber fruits can be stored for longer period at the temperature of

a) 5°C b) 10°C

c) 25°C d) None of these

37. Mucilaginous substance in cluster bean is called

a) Allantoin b) Allantonic acid

c) Glycoprotein **d) Mannogalacton**

38. Cluster bean is sown at a spacing of

a) 75 x 60 cm b) 60 x 45 cm

c) 45 x 20 cm d) None of these

39. Cluster bean is

a) Self-pollinated b) Often cross-pollinated

c) Cross pollinated d) None of these

40. Cluster bean produces an amide

a) Aspargine b) Glycoprotein

c) Allantoin d) Allantonic acid

41. Colour of Pusa Sweti variety of turnip is

a) Red b) Golden

c) White d) None of these

42. Conical head type of cabbage cultivars have. maturity

a) Early **b) Mid**

c) Late d) None of these

43. Contender is a variety of

a) French bean b) Lima bean

c) Broad bean d) None of these

44. Healthy and stocky seedlings of cauliflower are transplanted at a distance of

a) 60 x 30-45 cm b) 60 x 60 cm

c) 40 x 15 cm d) 45 x 45 cm

45. Hermaphrodite cultivar of ridge gourd is

a) Pusa Nasdar
b) Satputia
c) CO-I
d) Jhnga Torai

46. Heterosis in bottle gourd was reported in a number of crosses by

a) I.D. Tyagi (1973)
b) V. S. Seshadri (1988)
c) G. S. Gill (1980)
d) U.K. Verma (1985)

47. The most important chilly growing State is

a) Karnataka
b) Uttar Pradesh
c) Rajasthan
d) Andhra Pradesh

48. The most nutritive type of broccoli is

a) Green coloured
b) White coloured
c) Purple coloured
d) None of these

49. The most popular salad crops is

a) Lettuce
b) Parsley
c) Celery
d) None of these

50. The most popular variety of watermelon is

a) Ashi Yamato
b) Sugar Baby
c) Shiper
d) None of these

51. The most serious disease of okra is

a) Yellow vein mosaic
b) Powdery mildew
c) Root rot
d) None of these

52. The most serious virus disease of bean is

a) Common bean mosaic
b) Curly top
c) Yellow bean mosaic
d) None of these

53. The most serious virus disease of tomato in the rainy and autumn seasons is

a) TMV
b) CMV
c) Leaf curl
d) Tomato spotted wilt virus

54. The most suitable time for transplanting sweet pepper is

a) Mid-February
b) March-April
c) April-May
d) None of these

55. The most troublesome insect pest of lettuce is

a) Leaf miner b) Painted bug

c) Aphid d) None of these

56. The native place of garden beet is

a) Asia **b) Europe**

c) America d) None of these

57. High quality of pea seed should have a minimum germination percentage of:

a) 75 b) 95

c) 85 d) 65

58. Highest pod yield in French bean is obtained at a soil pH of

a) 5.3-6.0 b) 6.0-7.0

c) 7.5-8.5 d) None of these

59. Horse radish is propagated by

a) Seeds **b) Earthling**

c) Root cuttings d) All of the above

60. Hyacinth bean or sem originated in

a) India b) America

c) China d) None of these

61. Angoorlata is a popular variety of

a) Grapes b) Pea

c) Tomato d) Musk melon

62. Which of the following vegetable is perennial?

a) Asparagus b) Cauliflower

c) Garlic d) Indian bean

63. Which of the following is a major disease of onion in India?

a) Smut b) Leaf blight

c) White rot **d) Purple blotch**

64. 'Pusa Chetki' is a variety of

a) Radish b) Carrot

c) Turnip d) Spinach

65. Which of the following vegetable has highest amount of vitamin-A?

a) Amaranths
b) Lettuce
c) Carrot
d) Cauliflower

66. Pungency in radish is due to

a) Solanin
b) Allicin
c) Isothioctanate
d) Capsecin

67. 'Suverna' is a popular variety of

a) Coriander
b) Ginger
c) Turmeric
d) Fenugreek

68. Which of the following vegetable is dioecious in nature?

a) Brinjal
b) Bottle gourd
c) Pointed gourd
d) Pumpkin

69. Cytoplasmic male sterility is used for hybrid seed production in

a) Okra
b) Brinjal
c) Onion
d) Tomato

70. Which of the following is a salt tolerant crop?

a) Cowpea
b) Pea
c) Sugar beat
d) Musk melon

71. Yellow Vein Mosaic of Okra is transmitted by:

a) Aphid
b) Beetle
c) Wind
d) White fly

72. Powdery mildew disease is caused by

a) Bacteria
b) Virus
c) Fungus
d) None of these

73. "Buck eye rot" is associated with

a) Tomato
b) Brinjal
c) Pea
d) Sweet potato

74. "Pusa Meghali" is an improved variety of

a) Beet root
b) Radish
c) Carrot
d) Turnip

75. "Parbhani Kranti" is a popular variety of

a) Brinjal
b) Chilli
c) Potato
d) Okra

76. Which of the following crop is produced maximum in the world?

a) Tomato
b) Chilli
c) Potato
d) Cauliflower

77. "Browning" in cauliflower is due to

a) Bo deficiency
b) Zn deficiency
c) Mo deficiency
d) N deficiency

78. How much seed is required for one hectare of tomato crop

a) 1200-1500g
b) 400-500g
c) 800-1000g
d) 100-200g

79. Which rank India has in vegetable production in the world

a) First
b) Second
c) Third
d) Fourth

80. Which of the following is a cross-pollinated vegetable?

a) Chilli
b) Cauliflower
c) Tomato
d) All of these

81. Vegetables are mostly canned in

a) Syrup
b) Water
c) Brine
d) None of these

82. Seedless watermelons are

a) Diploid
b) Triploid
c) Hexaploid
d) Tetraploid

83. Leaf curl resistant tomato variety is:

a) Sioux
b) Punjab Chhuhara
c) Hisar Anmol
d) S-12

84. Polyploidy method of improvement can be followed for

a) Cucurbits
b) Leafy vegetables
c) Cole crops
d) None of these

85. Parthenocarpy does not pose a major problem in

a) Corriander
b) Leafy vegetables
c) Cucumber
d) All of the above

86. Hybrid vegetable grows maximum in

a) Brinjal
b) Cabbage
c) Tomato
d) None of these

87. Pusa Chetki is a variety of

a) Radish

b) Cabbage

c) Cauliflower

d) Tomato

88. Pusa Jyothi is an improved variety of

a) Musk melon

b) Cowpea

c) Palak

d) Bottle gourd

89. Protein percentage in pea is

a) 5.1 per cent

b) 8.3 per cent

c) 7.2 per cent

d) 9.0 per cent

90. 'Petha' is prepared from

a) Ridge gourd

b) Ash gourd

c) Pointed gourd

d) Bottle gourd

91. Edible part of asparagus is known as

a) Spears

b) Bud

c) Stem

d) Seed

92. Odour of garlic is due to

a) Allyl di-sulphate

b) Diallyl di-sulphhide

c) Quercetin

d) None of these

93. Odour of onion is due to

a) Allyl di-sulphide

b) Propyl di-sulphide

c) Diallyl di-sulphide

d) Allyl propyl di-sulphide

94. Odour of radish is due to

a) Quercetin

b) Isothiocyanate

c) Allyl di-sulphide

d) Sinigrin

95. Odour of cole crops is due to

a) Sinigrin

b) Quercetin

c) Capsecin

d) Marmelocin

96. Sweet potato is propagated by

a) Tubers

b) Root Cutting

c) Runner

d) Stem cutting

97. Pusa Meghna is a variety of

a) Cow pea

b) Watermelon

c) Carrot

d) Cauliflower

98. Edible part of rat tail radish is

a) Roots
b) Stem
c) Pods
d) Tuber

99. Acid tolerant vegetable is

a) Onion
b) Broccoli
c) Lettuce
d) Rhubarb

100. Pusa Lal is a variety of

a) Carrot
b) Pumpkin
c) Onion
d) Sweet potato

101. Pusa Red is a variety of

a) Onion
b) Carrot
c) Sweet potato
d) None of these

102. Edible part of Jerusalem artichoke is

a) Flower bud
b) Stem
c) Tuber
d) Seeds

103. Parent of Pusa Sanjog is

a) Japanese Gynoecious lines x Green Long Naples
b) Japanese Long Green x Himangi
c) Poinsett x Green Long Naples
d) Sheetal Himangi

104. Pusa Sanyog is a hybrid of

a) Ash gourd
b) Bitter gourd
c) Cucumber
d) Ridge gourd

105. Sex form in spinach is

a) Heteromorphic
b) Dioecious
c) Apomixis
d) Tetramorphic

106. Catface is a physiological disorder of

a) Potato
b) Tomato
c) Brinjal
d) Cauliflower

107. Zening is a physiological disorder of

a) Beet
b) Artichoke
c) Carrot
d) Coriander

108. Buttoning is a physiological disorder of

a) Button mushroom **b) Cauliflower**

c) Tomato d) Potato

109. Draught resistant vegetable is

a) Tomato b) Pea

c) Sweet potato d) Cauliflower

110. White brinjal is good for

a) Cancer b) Heart diseases

c) Diabetes d) Headache

111. Based on the method of culture vegetable are classified into

a) 7 groups **b) 13 groups**

c) 15 groups d) 10 groups

112. Carbohydrate content in potato (in 100 g) is

a) 15.4 g b) 17.8 g

c) 22-6 g d) 31-6 g

113. Frost resistant variety of potato is

a) Kufri Sinduri b) Kufri Alanka

c) Kufri Jyoti **d) Kufri Sheetman**

114. Kufri Naveen is resistant to

a) Wart b) High temperature

c) Low temperature d) Salt tolerance

115. Sprouting of potato can be controlled by a pre harvest spray of

a) IBA **b) MH**

c) IAA d) None of these

116. Soft rot of vegetable is caused by

a) Alternaria solani b) Seetoria lycopersici

c) Erwinia Carotovora d) None of these

117. Cu requiring vegetable is

a) Onion b) Potato

c) Tomato d) Cauliflower

118. Storage life of onion is reduced due to deficiency of

a) Potash b) Calcium

c) Nitrogen d) Phosphorus

119. Single seeded cucurbit is

a) Chow Chow b) Kakrol

c) Pointed gourd d) Ivy gourd

120. ICMR recommended per capita consumption of vegetable in a day

a) 175 g b) 285 g

c) 250 g **d) 300 g**

121. Potato disease caused by alkaline condition is

a) Black scurf **b) Scab**

c) Late blight d) None of these

122. Tuber production in potato stops at

a) 10°C **b) 29 °C**

c) 12°C d) 15°C

123. Purple Vienna is a variety of

a) Beet b) Brinjal

c) Radish **d) Khol Khol**

124. Riceyness resistant vareity of cauliflower is

a) Pant Shubra b) Improved Japanese

c) Pusa Subhra d) Pusa Deepali

125. Suitable variety of pea for dehydration is

a) Salvia b) Lincoln

c) Arkel **d) Bonneville**

126. Pusa Jwala variety of chilli is resistant to

a) Fungal disease b) Nematode

c) Viral disease d) Insect-pest

127. Pusa Manjari is a hybrid of

a) Bottle gourd b) Baby corn

c) Cowpea d) Brinjal

128. Optimum temperature for onion bulb formation is

a) 10-12°C b) 25-28°C

c) 15-18°C **d) 20-22°C**

129. Multiple disease resistant variety of water melon is

a) Arka Jyoti b) Arka Bedana

c) Arka Manik d) All of these

130. Bitter principle in bitter gourd is

a) Momerdicin
b) Capsacin
c) Quercetin
d) None of these

131. Bitter principle in Luffa is

a) Momordicin
b) Capsacin
c) Quercetin
d) Luffein

132. To induce flowering growth in sweet potato regulator is used

a) IBA
b) GA3
c) ABA
d) 2,4-D

133. Daily requirement of root and tuber vegetable per head is

a) 75 g
b) 125 g
c) 100 g
d) 300 g

134. High temperature condition for fruit setting tomato variety is

a) Pusa Early dwarf
b) Pusa Hybrid-1
c) Pusa Uphar
d) None of these

135. High and low temperature for fruit setting tomato variety is

a) Pusa Sheetal
b) Pusa Sadabahar
c) Pusa Gaurav
d) Pusa Ruby

136. Dioecious vegetable is

a) Aspalagus
b) Ivy gourd
c) Pointed gourd
d) All of these

137. Predominant sex form in musk melon is

a) Gynoceous
b) Andromanoecious
c) Monoceous
d) None of these

138. Bush type cucurbit vegetable is

a) Ivy gourd
b) Summer squash
c) Bottle gourd
d) Bitter gourd

139. Anemophilous vegetable is

a) Amaranth
b) Spinach
c) Beet
d) All of these

140. Marsh spot in pea is due to

a) Mn deficiency
b) Zn deficiency
c) Cu deficiency
d) Ca deficiency

141. Snowball is a popular variety of

a) Cabbage
b) Brinjal
c) Tomato
d) Cauliflower

142. 'Leaf curl' disease in Chilli is transmitted through

a) Thrips
b) Hopper
c) Aphid
d) White fly

143. Which of the following is not matched correctly?

a) Pusa Sadabahar - Tomato
b) Pusa Jyoti - Watermelon
c) Pusa Naveen - Bitter gourd
d) Pusa Sawani - Okra

144. Tomato is a rich source of

a) Riboflavin
b) Carotene
c) Xanthophyll
d) None of these

145. 'Pant Rituraj' is an improved cultivar of

a) Carrot
b) Okra
c) Tomato
d) Brinjal

146. Which of the following is the native place of Turmeric

a) Peru
b) China
c) Malaysia
d) India

147. In which of the following vegetable first hybrid was released?

a) Bottle gourd
b) Chilli
c) Tomato
d) Cowpea

148. First intergenic hybrid between radish and cabbage was made by

a) Karpechenko
b) Hull
c) Rimpu
d) Andrew Night

149. Pusa Sadabahar is a variety of

a) Tomato
b) Chilli
c) a and b
d) None of these

150. Arka Gaurav is an improved cultivar of

a) Tomato
b) Capsicum
c) Brinjal
d) Cabbage

151. Which one of the following countries is having the first rank in area and production of Garlic?

a) India
b) Burma
c) China
d) Japan

152. Carrot is a good source of

a) Fat
b) Lycopene
c) Protein
d) Carotene

153. *Solanum tuberosum* is a

a) Diploid
b) Haploid
c) Tetraploid
d) None of these

154. Vegetables are known as protected food because they are good source of

a) Vitamins and Minerals
b) Carbohydrates
c) Fat
d) All of these

155. Garlic flavour is due to presence of

a) Dimethyl trisulphide
b) Allylpropyl disulphide
c) Diallyl disulphide
d) Propyl disulphide

156. Toxic substances 'Saponine' is present in

a) Celery
b) Potato
c) Brinjal
d) Spinach

157. 'Dioscorine' a toxic substance is present in

a) Yams
b) Potato
c) Cassava
d) Beans

158. Which of the following factor affects the storage life of vegetables?

a) Maturity stage at harvesting
b) Temperature of storage
c) Humidity of storage
d) All of the above

159. Which is/are physiological disorders caused by Boron deficiency?

a) Brown heart of beet root
b) Splitting of carrot
c) Browning of cauliflower
d) All of the above

160. Bitterness in carrot occurs due to

a) Excessive moisture
b) Excessive exposure to sunlight
c) Deleterious effect of ethylene
d) Poor ventilation

161. Which of the following variety of potato is not resistant to wart disease?

a) Kufri Muthu
b) Kufri Jyoti
c) Kufri Sinduri
d) Kufri Sherpa

162. Which of the following vegetable is not a C3 crop?

a) Sweet potato
b) Tomato
c) Amaranth
d) Lettuce

163. Which of the following is not a cross pollinated crop?

a) Lettuce
b) Ash gourd
c) Onion
d) Spinach

164. Which is the richest source of carbohydrates?

a) Spinach
b) Tapioca
c) Sweet potato
d) Potato

165. Black heart is a physiological disorder of

a) Potato
b) Tomato
c) Cabbage
d) Capsicum

166. In tomato, blossom end rot is caused due to deficiency of

a) P
b) Ca
c) Mg
d) B

167. Pusa Meghali is an improved variety of

a) Onion
b) Carrot
c) Tomato
d) Bottle gourd

168. Dolichos bean originated from

a) South America
b) Central America
c) India
d) Europe

169. "Pusa Madhuras" is a variety of

a) Musk melon
b) Water melon
c) Tomato
d) Spong gourd

170. Garlic are harvested in the month of

a) July-August
b) March-April
c) November-December
d) None of these

171. Seed rate (g/hectare) of Brinjal is

a) 150-300 g
b) 375-500 g
c) 500-750 g
d) 700-1000 g

172. The edible portion of onion is

a) Bulb b) Knob
c) Head d) Crown

173. Little leaf in Brinjal is due to

a) Mycoplasma b) Virus
c) Fungus d) Bacteria

174. Which one is the cultivar of bottle gourd?

a) Pusa Meghdut b) Pusa Green
c) Pusa Vishvas d) Australian Green

175. In how many days late variety of cabbage is harvested?

a) 60-70 days b) 80-90 days
c) 100-120 days d) 110-140 days

176. Seed rate of early season cauliflower per hectare is

a) 100-200 g b) 700-1000 g
c) 450-700 g d) 300-400 g

177. Onion seedlings are transplanted at the age of

a) 1-2 weeks b) 3-4 weeks
c) 6-7 weeks d) 9-10 weeks

178. Higher temperature at the time of flowering in bottle gourd induces

a) More male flowers b) More female flowers
c) More bisexual flowers d) None of these

179. The best suited soil pH range for tomato cultivation is

a) 5.0-6.0 b) 6.0-7.0
c) 7.0-8.5 d) 8.5-9.0

180. The soil of nursery bed should be sterilized by

a) Dimecron b) BHC
c) Formaldehyde d) Thiram

181. Which of the following is correctly matched?

a) Potato - Corm b) Radish - Conical
c) Carrot - Fusiform **d) Beet - Napiform**

182. Which of the following soil is best suitable for vegetable production?

a) Sandy b) Clay
c) Sandy Loam d) Clay Loam

183. Advance form of sex in cucurbits is

a) Andromonoeciou b) Dioecious

c) Hermaphrodite d) Monoecious

184. 'Root bulging' is common process in

a) Carrot **b) Radish**

c) Onion d) Sugarbeet

185. 'Inbreeding depression' is found in

a) Bottle gourd and Brinjal b) Musk melon and Cucumber

c) Tomato and Chillies **d) Onion and Cabbage**

186. 'Arka Nishant' is a popular cultivar of

a) Carrot **b) Radish**

c) Onion d) Garlic

187. Which of the following is/are the celery variety?

a) Standard Beared b) Wright Grove Giant

c) Fork Hook Emperor **d) All of the above**

188. 'Bountiful' is a variety of

a) Cowpea **b) French bean**

c) Lettuce d) Broccoli

189. Single gene male sterility is transferred in .... Years

a) 3-4 **b) 5-6**

c) 7-8 d) 4-5

190. Triploid watermelon variety is

a) New Hampshire Midget b) Charleston Grey

c) Pusa Bedana d) None of these

191. ***Trigonella corniculata*** **produces flowers of following colour**

a) White b) Red

c) Purple d) Yellow

192. Major pollinating agent in cucurbits is

a) Ants **b) Honey bee**

c) House fly d) Bumble bee

193. Tuberization in sweet potato is promoted under

a) 7 hr. light b) 9 hr. light

c) 11 hr. light d) 13 hr. light

194. The most common cucurbit grown under glasshouses

a) Bottle gourd b) Water melon
c) Musk melon d) Cucumber

195. Potato tuber production is maximum at following temperature in celcious

a) 15 **b) 20**
c) 25 d) 30

196. Indian spinach or Poi belongs to the family

a) Chenopodiaceae b) Leguminosae
c) Basellaceae d) Convolvulaceae

197. Pollination in beetroot is by

a) Insect b) Water
c) Wind d) All of the above

198. Rat tail radish is grown for its

a) Long thin pod b) Tender roots
c) Shoots d) Stem

199. Pusa Rasraj is a variety of

a) Long melon **b) Musk melon**
c) Water melon d) None of these

200. Salt-tolerant variety of onion

a) N-53 b) Nasik red
c) N-404 **d) Punjab selection**

201. In turnip pithiness resistant variety is

a) Pusa Kanchan b) Pusa Swati
c) Pusa Swarnima d) Pusa Chandrima

202. Multigerm seed is found in case of

a) Onion b) Carrot
c) Beetroot d) Tomato

203. The country where maximum genetic diversity in brinjal occurs is

a) India b) Peru
c) China d) Iran

204. Isolation distance for certified seed production in radish is

a) 800m **b) 1000m**

c) 1600m d) 200m

205. Onion seeds remain viable for germination for following years

a) 1-2 years **b) 3-4 years**

c) 2-3 years d) 4-5 years

206. Little leaf disease of brinjal is spread by

a) Aphid **b) Leaf-hopper**

c) Whitefly d) None of these

207. Which one of the following vegetable crop can tolerate acid soils

a) Cucumber **b) Water melon**

c) Okra d) Beans

208. Pinching of male buds before anthesis in the female parent in cucurbits was first suggest by

a) Choudhary and Singh b) Seshadri and More

c) Tyagi and Singh d) V swaroop

209. In green vegetable,the limiting amino acids is

a) Arginine b) Lysine

c) Methionine d) Tryptophan

210. While Arka Pragati is an onion; Pusa Pragati is

a) Cluster bean **b) Garden pea**

c) Broad bean d) Lima bean

211. Phenolic factor present in onion which has antifungal properties

a) Sinigrin b) Quercetin

c) Catechol d) Allylpropyl

212. How many types of gametes will be produced by F2 in a dihybrid cross

a) 4 b) 2

c) 9 d) 16

213. Potato contains how much per cent of protein

a) 1.8 **b) 1.4**

c) 2.2 d) 3.2

214. The stalk of the ovule is called

a) Pedicel
b) Panicle
c) Petiole
d) Micropyle

215. Bush pepper is propagated from

a) Laterals
b) Hanging shoots
c) Top shoots
d) Runners

216. TSS of watermelon is

a) 10-12 ^{0}Brix
b) 14-16^{0}Brix
c) 18-22^{0}Brix
d) 20-22^{0}Brix

217. Essential oil per cent in dill is

a) 0.1-2.6 per cent
b) 0.1-1.0 per cent
c) 1.5-4.0 per cent
d) None of these

218. Commercially used growth retardant is

a) Paclobutrazol
b) BA
c) GA3
d) None of these

219. Capacity of refrigeration is measured by

a) Gram
b) Pascal
c) Lux
d) Ton

220. Bottleneck is a disorder of

a) Cucumber
b) Carrot
c) Bottel gourd
d) None of these

221. Sree Dhara is a variety of

a) Sweet potato
b) Cassava
c) Coleus
d) None of these

222. Polyphenol oxidase enzyme inactivate at the temperature of

a) 80-90 ^{0}C
b) 50^{0}C
c) 100^{0}C
d) Above 100^{0}C

223. Heterosis in bottle gourd is reported by

a) ID Tyagi
b) GS Gill
c) VS Seshadri
d) UK Verma

224. The hybrids crops most preferred in India is

a) Brinjal
b) Onion
c) Cucumber
d) Tomato

225. Drought-tolerant onion is

a) Hissar2

b) Pusa Ratnar

c) Pusa Red

d) Baswant-780

226. In case of chilli perennial variety is

a) Pusa Jwala

b) Arka Lohit

c) Pusa sadabahar

d) Arka Abir

227. Man-made vegetable

a) Rutabaga

b) Winged bean

c) Cluster bean

d) Artichoke

228. Homozygosity and homogeneity is max. in

a) Inbreed line

b) Land races

c) Pure line

d) Synthetic varieties

229. Downy mildew of muskmelon Is controlled by systemic fungicide

a) Ridomil MZ

b) Blitox-50

c) Indofil M45

d) None of these

230. Delphinidin a part pigment is responsible for development of ______ colour

a) Pink

b) Red

c) Orange

d) Blue

231. Pionsette is the popular variety of

a) Long melon

b) Cucumber

c) Water melon

d) Musk melon

232. Which is anti-gibberellin factor

a) cycocel

b) cultar

c) BA

d) IBA

233. Which one of the following is used for production of virus free plants

a) Anther

b) Nodes

c) Meristems

d) Ovule

234. Which cucurbit have highest vit-A

a) Pumpkin

b) Cucumber

c) Bitter gourd

d) All of these

235. Artificial pollination is essential in

a) Cinnamon
b) Vanilla
c) Clove
d) Nutmeg

236. Plug trays are related to

a) Vegetable harvesting
b) Fruit harvesting
c) Nursery management
d) None of these

237. Sex form present in betel vine

a) Monoecious
b) Dioecious
c) Andromonoecious
d) None of these

238. Poisonous baiting is done to control

a) Borer
b) Fruit-fly
c) Thrips
d) Whitefly

239. Florida butter is an important variety of

a) Broad bean
b) Lima bean
c) Indian bean
d) Winged bean

240. Use of rootstock is common in

a) Cole crop
b) Cucurbits
c) Tuber crops
d) Leafy vegetables

241. Arka Alok is a tomato variety resistant to

a) Wilt
b) Nematode
c) Fruit borer
d) None of these

242. ______is used as donor parent for salt tolerance in tomato breeding

a) L. esculentum
b) L. hirsutum
c) L. cheesmanii
d) L. pimpinellifolium

243. Generally the frequency of spontaneous mutation is

a) 10^-6
b) 10^-4
c) 10^-5
d) 10^-7

244. Which type of mutation is used in crop improvement

a) Vital
b) Lethal
c) Sub-lethal
d) Sub-vital

245. A cross between an inbreed line and an open pollinated variety is called as

a) Test cross
b) Top cross
c) Back cross
d) Poly cross

246. Spontaneous mutation used mostly for

a) Ornamental
b) Spices
c) Fruits
d) Vegetables

247. Select the element constituent 0f amino acid which is essential for protein synthesis

a) Calcium
b) Magnesium
c) Manganese
d) Sulphur

248. A nullisomic Individual is represented by

a) 2n-1
b) 2n-2
c) 2n+1
d) 2n+2

249. In heterosis breeding one should capitalize on

a) Inbreeding depression
b) Over dominance
c) Homozygosity
d) High seed cost

250. Which of the following tomato variety has the indeterminate growth habit

a) Sioux
b) Roma
c) Pusa gaurav
d) Pusa Early Dwarf

251. Many fold effects of a gene refers to

a) Penetrance
b) Epistasis
c) Pleiotrophy
d) None of these

252. Gherkin, *Cucumis anguria* is a mutant of

a) C. moschata
b) C. pepo
c) C. longipes
d) C. sativus

253. Which of the following carrot variety is introduced from Germany and suitable for Nilgiri Hills

a) Imperator
b) Zeno
c) Chanteny
d) Danvers

254. Metsubre is a disorder of ______ due to calcium deficiency

a) Colocasia
b) Yam
c) Elephant foot yam
d) Potato

255. Which of the following is a flower type of brinjal

a) Long styled b) Short-style
c) Pseudo-short style **d) All of the above**

256. Frog eye rot is a physiological disorder of

a) Chilli b) Potato
c) Tomato d) Okra

257. Yerrusseri product is prepared from which vegetable

a) Cucumber b) Carrot
c) Pumpkin d) None of these

258. Starch content per cent in sweet potato tuber is

a) 16 b) 6
c) 26 d) 0.6

259. Optimum seed rate (g/ha.) of cauliflower

a) 200-300 **b) 300-500**
c) 700-800 d) 800-900

260. Absence of axillary bud is the typical feature of

a) Cauliflower b) Cabbage
c) Brussels sprout d) None of these

261. Sponge gourd and ridge gourd are

a) Intercrossable b) Intersterile
c) Both d) None of these

262. Which of the following cucurbits exhibits vivipary

a) Cucumber **b) Chow-chow**
c) Pumpkin d) Musk melon

263. Arka Jyoti is a watermelon while Pusa Jyoti is

a) Palak b) Beetroot
c) Amaranthus d) Sugarbeet

264. The yellow colour in tomato is due to pigment

a) Anthocyanin b) Carotene
c) Lycopene **d) Pro-lycopene**

265. Arka Jeet is a popular variety of

a) Musk melon b) Water melon
c) Pumpkin d) Bitter gourd

266. Symptoms of calacium deficiency in plants is first noticed in

a) Middle leaves
b) Terminal leaves
c) Lower leaves
d) None of these

267. Ploidy level of cultivated potato is

a) Diploid
b) Tetraploid
c) Triploid
d) Hexaploid

268. While Pusa Gaurav is tomato, Arka Gaurav is

a) Tomato
b) Brinjal
c) Capsicum
d) Chilli

269. Brown Heart is a physiological disorder of

a) Carrot
b) Turnip
c) Radish
d) Beet

270. Pant Samrat is brinjal while samrat is

a) Potato
b) Sweet potato
c) Tapioca
d) Colocasia

271. Pusa Bindu is brinjal while Arka Bindu is

a) Tomato
b) Onion
c) Cucumber
d) Colocasia

272. Yellow anther type variety in chilli

a) Bhagya lakshmi
b) Andhra jyoti
c) Jwala mukhi
d) Bhaskar

273. Haploid plants have been produced by anther culture in tomato by

a) Greshof and Doy
b) Mehta and Upadhyay
c) Tyagi and Prasad
d) H.B. Singh

274. Netted variety of Muskmelon are known as

a) Honey Dews
b) Casabas
c) Cantaloupes
d) None of these

275. The crop with which turnip crosses easily is

a) Radish
b) Mustard
c) Cabbage
d) None of these

276. In metaphase stage of cell division chromosome are arranged on

a) Centromere
b) Equitorial plate
c) Polar plate
d) None of these

277. Quantitative characters are governed by

a) Monogene b) Bi gene

c) Recessive gene **d) Polygene**

278. Arka Abir is the variety of

a) Chilli b) Tomato

c) Brinjal d) Radish

279. *Sauropus androgynous*, chekurmanis belongs to the family

a) Euphorbiaceae b) Convolvulaceae

c) Solanaceae d) None of these

280. Which type of tomato are referred as self-topping

a) Determinate b) Indeterminate

c) Intermediate d) None of these

281. Growth regulator which improves fruit set in tomato is

a) MH b) GA3

c) Ethephone **d) 2,4-D**

282. A phenomenon in which a gene has more than one phenotypic effect is

a) Polymorphism **b) Pleiotropism**

c) Penetrance d) Expressively

283. Which of the following is a major disease of onion in India

a) Leaf blight **b) Purple blotch**

c) Smut d) White rot

284. Botanical name of pumpkin is

a) C. moschata b) C. maxima

c) *C. ficifolia* d) None of these

285. Botanical name of radish is

a) Daucus carota b) Brasica oliracea

c) Raphanus sativus d) Allium cepa

286. White Brinjal is a good remedy for

a) Diabetes b) Heart disease

c) Cancer d) Skin problem

287. Per capita consumption of potato in India is

a) 15 kg b) 20 kg

c) 30kg d) 50 kg

288. The variety resistant to pithiness in turnip is

a) Pusa Swati
b) Pusa Chandrima
c) Pusa Swarnima
d) Pusa Kanchan

289. The optimum temperature requirement for snowball group of curd initiation and development in cauliflower is

a) 20-27 ^{0}C
b) 20-25 ^{0}C
c) 12-16 ^{0}C
d) 10-16 ^{0}C

290. The first tropical variety which can set seeds in in plains in cabbage is

a) Pusa Sambandh
b) Pusa Ageti
c) Pusa drum-head
d) Golden acre

291. The variety is most suitable for export of onion is

a) Arka Bindu
b) Arka Kalyan
c) Pusa Red
d) N-53

292. Akashin is disorder in radish is due to deficiency of

a) Mn
b) B
c) Ca
d) Mg

293. Jadecross is a variety of

a) Brussels sprout
b) Sprouting Broccoli
c) Knol-khol
d) Cabbage

294. Chieftain is a variety of

a) Capsicum
b) Brussels sprout
c) Broccoli
d) Savoy Cabbage

295. King of North is variety of

a) Cabbage
b) Knol-khol
c) Cauliflower
d) Brussels sprout

296. Cucumber green mottle virus is transmitted through

a) Insect
b) Soil
c) Water
d) Air

297. Balsam pear is common name of

a) Bottle gourd
b) Cucumber
c) Bitter gourd
d) Pointed gourd

298. Botanical name of scarlet gourd is

a) Momordica charantia

b) Tricosanthus dioica

c) Coccinia indica

d) None of these

299. Onion variety suitable for dehydration is

a) Pusa Red

b) Pusa Ratnar

c) Pusa White Flate

d) Pusa White Round

300. *Colletrotrichum capsici* causes which diseases in chilli

a) Anthracnose

b) Fruit Rot

c) Dieback

d) All of these

301. Ridge gourd is commercially trained through

a) Bower system

b) Angular system

c) Kniffin system

d) None of these

302. Money maker is the variety of

a) Tomato

b) Brinjal

c) Chilli

d) Potato

303. Seed Plot Technique (SPT) in potato was developed by

a) G Kallo

b) Harbhajan Singh

c) Pushkarnath

d) None of these

304. Which is a vegetable cum spice crop

a) Fenugreek

b) Cumin

c) Coriander

d) Curry leaf

305. Dry chilli contain how much per cent of pericarp

a) 10

b) 25

c) 30

d) 40

306. The fruit of Okra botanically known as

a) Pod

b) Berry

c) Capsule

d) Silique

307. The inflorescence of Onion is called as

a) Umbell

b) Spadix

c) Panicle

d) Raceme

Chapter 14

MCQ in Postharvest Management

1. **Deep bed drying is characterized by**
   a) Ambient temperature drying
   b) Uniform drying rate
   c) Homogenous drying
   d) Heterogeneous drying

2. **Screw conveyor is generally used to move grains horizontally but it can be also used up to an angle of**
   a) 15^0
   b) 30^0
   c) 45^0
   d) 90^0

3. **The level of post-harvest losses in fruits and vegetables is**
   a) 10-15 per cent
   b) 25-35 per cent
   c) 20-25 per cent
   d) 35-40 per cent

4. **Treating the flowers with high concentrations of silver nitrate and nickel for few time is usually done in**
   a) Impregnation
   b) Pulsing
   c) Bud opening
   d) Pre-cooling

5. **In a continuous flow dryer, dry air temperature is**
   a) 70 ^{0}C
   b) 80 ^{0}C
   c) 65 ^{0}C
   d) None of these

6. **Due to improper handling of fruits and vegetables the total waste is**
   a) 2 per cent
   b) 5 per cent
   c) 10 per cent
   d) 40 per cent

7. CA stands for

a) Centrally air-conditioned
b) Completely air-conditioned
c) Controlled atmosphere
d) None of these

8. The FPO in India follows the food safety and standards act of

a) 2005
b) 2006
c) 2007
d) 2008

9. The FPO is effective from

a) 1955
b) 1957
c) 1956
d) 1961

10. The certifying agency of FPO is

a) Ministry of food processing industries
b) Ministry of fruit processing industries
c) Ministry of food protection industries
d) Ministry of food production industries

11. Central food laboratory is situated at

a) Bangalore
b) Hyderabad
c) New Delhi
d) Kolkata

12. Central food technological research institute is situated at

a) Chennai
b) **Mysore**
c) Goa
d) Thiruvananthapuram

13. CIPHET was first established in

a) 1979, Abohar
b) 1989, Ludhiana
c) 1978, Gurudaspur
d) 1987, Amritsar

14. IIVR situated at

a) Kolkata
b) Chennai
c) Patna
d) Varanasi

15. E224 is

a) KMS
b) TBZ
c) Sodium Meta-bisulphide
d) None of these

16. CEO of FSSAI is

a) Dr. M. R. Krishnamurthy
b) Arun Singhal
c) Smt. Jyotsna Pal
d) Dr. Yash Parmar

17. MVTR stands for

a) Moisture vapour transmission rate

b) Moisture vapour transmittance rate

c) Moisture vapour transition rate

d) None of above

18. Chairperson of FSSAI is

a) M. Shyamla | b) Girish Randhawa

c) Rajesh Bhushan | d) R. T. Shekhawat

19. ZECC is also known as

a) Evaporative cooling chamber | b) No cost cool chamber

c) Underground cool chamber | d) All of above

20. ZECC was developed by

a) Shrinivasan and V.V. Reddy | **b) S. K. Roy and D.S. Khurdiya**

c) G.L. Sharma and S. Krishna | d) None of the above

21. Which is the mode of deterioration during storage is

a) Fiber development | b) Root growth

c) Loss of moisture | **d) All of above**

22. Crop production factors responsible for post-harvest losses is

a) Curing | b) Light

c) Maturity | **d) All of above**

23. Storage and transport conditions responsible for post-harvest losses is

a) Atmospheric gases | b) Packaging

c) Temperature | **d) All of above**

24. Production capacity of small scale food processing industry is

a) 10-15 t/year | b) 15-25 t/year

c) 50-150 t/year | **d) 50-250 t/year**

25. Production capacity of cottage scale food processing industry is

a) 10-15 t/year | b) 10-50 t/year

c) 50-150 t/year | d) 50-100 t/year

26. Production capacity of large-scale food processing industry is

a) 10-15 t/year | b) 15-25 t/year

c) 200-250 t/year | **d) More than 250 t/year**

27. Major importer of fruit juice and jam from India is

a) Sri Lanka
b) China
c) England
d) Russia

28. Major importer of canned fruit from India is

a) RSA
b) Pakistan
c) UAE
d) Australia

29. Major importer of canned vegetables from India is

a) UK
b) UAE
c) USA
d) All of above

30. Major importer of dehydrated onion from India is

a) Japan
b) Korea
c) China
d) All of above

31. Major importer of fruit pickles from India is

a) USA
b) New Zealand
c) Sri Lanka
d) None of above

32. APEDA replaced by

a) FPO
b) FSSAI
c) FAA
d) PFEPC

33. Deglet Nour is

a) Coconut
b) Date palm
c) Peanut
d) Walnut

34. A daily meal should include per capita is

a) 100g of tuber vegetable
b) 200g of tuber vegetable
c) 250g of tuber vegetable
d) 300g of tuber vegetable

35. A daily requirement of carbohydrates is

a) 250-300g
b) 300-400g
c) 400-500g
d) 700-800g

36. Daily requirement of protein is

a) 40-50g
b) 50-60g
c) 60-70g
d) 70-80g

37. Water soluble vitamin is

a) Vitamin C
b) Vitamin B complex
c) Both a and b
d) All of above

38. Major source of B1 (thiamin) and Vitamin B2 (riboflavin) is

a) Brinjal
b) Leek
c) Okra
d) Palak

39. Daily requirement of vitamin c is

a) 50mg
b) 65mg
c) 70mg
d) 84mg

40. Per day per capita energy requirement for an adult is

a) 2100 kcal
b) 2300kcal
c) 2400kcal
d) 2600kcal

41. Which is a chemical change in food

a) Flavor
b) Color
c) Minerals
d) All the above

42. Which is a not type of changes

a) Microbial changes
b) Physical changes
c) Chemical changes
d) Mechanical changes

43. Principles of preservation is

a) Prevention of microbial decomposition
b) Prevention of self-decomposition
c) Prevention of damages by insect-pests
d) All of the above

44. Asepsis means

a) Physical killing of microbes
b) Chemical killing of microbes
c) Preventing entry of microbes
d) None of the above

45. Which is responsible for softening of the fruit structure

a) Amylase
b) Phenolase
c) Lipolytic
d) Proteolytic

46. Which group of microorganisms converts sugar into alcohol

a) Bacteria
b) Yeasts
c) Molds
d) All of the above

47. Lye peeling is generally done at temperature

a) 60 °C
b) 93 °C
c) 85 °C
d) None of the above

48. What is the critical point

a) Dried product becomes soft

b) Dried products become hard

c) Dried products become lumpy

d) None of the above

49. If the reading of the jelmeter is 3/4th, how much sugar would be added in jelly making

a) 500g | **b) 750g**

c) 1000g | d) 1250g

50. Which is considered as color

a) Black | b) Yellow

c) Orange | **d) Green**

51. Acid content in nectar is

a) 0.3 per cent | b) 0.4 per cent

c) 0.5 per cent | d) None of the above

52. Blackaning is the major problem of

a) Tuberose | b) Sauce

c) Apple | **d) Pickle**

53. What is the commercial method for carrot peeling

a) Hand peeling | b) Mechanical peeling

c) Lye peeling | **d) Steam peeling**

54. Presence of red color in jelly is due to

a) Pectin | b) Acid

c) Charring sugar | d) None of the above

55. Carbohydrate content in potato is

a) 22 per cent | b) 25 per cent

c) 30 per cent | d) 40 per cent

56. Chilling injuries arising from the exposure of products at a temperature

a) Above the normal physiological range

b) In CA storage

c) Below the normal physiological change

d) Under the poor ventilation condition

57. The volatile toxic substances accumulate in the tissue when fruit are stored in an environment of

a) Above the normal physiological range

b) In CA storage

c) Below the normal physiological change

d) Under the poor ventilation condition

58. What is the form of membrane lipids in fruits and vegetables that are resistant to chilling

a) Solid
b) Rigid
c) Fluid
d) Semi-fluid

59. Fresh fruits and vegetables such as apples, oranges and carrots are stored at

a) Above freezing
b) Below freezing
c) At freezing
d) 2°C

60. Surface pitting is the characteristics chilling injury in

a) Banana
b) Apple
c) Citrus
d) Pineapple

61. Internal discoloration is a common symptom of chilling injury in

a) Banana
b) Apple
c) Citrus
d) Pineapple

62. Fruit tissues pH is

a) Neutral
b) $^{3}5$
c) £5
d) $^{3}7$

63. Organic acid present in apple is

a) Citric acid
b) Tartaric acid
c) Malic acid
d) Benzoic acid

64. Different kinds of fruits can be stored together when their temperature requirement is

a) Should not same
b) Should be same
c) May be different
d) None of the above

65. Pre-cooling of fruits and vegetables are done at temperature

a) 5-10 °C
b) 10-12 °C
c) 15-17 °C
d) 15-20 °C

66. Which antioxidant is abundant in citrus fruit

a) Flavonoid
b) Tocopherol
c) Carotenoid
d) Ascorbic acid

67. Lycopene present in

a) Mango
b) Cucumber
c) Citrus
d) Tomato

68. Fruits and vegetables become brownish due to the presence of enzyme

a) Amylase
b) Protease
c) Lipo-oxidase
d) Polyphenol oxidase

69. The enzyme which produces bad smelling aldehydes in vegetables

a) Amylase
b) Protease
c) Lipo-oxidase
d) Polyphenol oxidase

70. Microbiological and nutritional shelf-life of minimally processed foods should be preferable up to

a) 7 day
b) 10 days
c) 21 days
d) 30 days

71. The quality problem in sliced potato and apple is due to

a) Putrefaction
b) Hydrolytic rancidity
c) Lypolytic rancidity
d) Enzymatic browning

72. Enzymatic browning in fruits and vegetables is due to

a) Enzyme
b) Oxygen
c) Copper and substarte
d) All of the above

73. The purpose of edible coating in minimally processed fruit and vegetables are

a) Sealing volatile flavor
b) Lower respiration
c) Retard ethylene production
d) All of the above

74. If in MAP the oxygen level is high then the temperature should be a range of

a) 0-3 ^{0}C
b) 5-10 ^{0}C
c) 10-15 ^{0}C
d) <20 ^{0}C

75. HP processing consist

a) High performance
b) High pasteurization
c) High pressure
d) None of the above

76. Brown heart disorder in apple due to the accumulation of toxic substances if

a) CO_2 is reduced during storage

b) CO_2 is increased during storage

c) O_2 supply is increased during storage

d) O_2 supply is reduced during storage

77. Banana, grape, fig are more susceptible of black mold rots due to

a) *Aspergillus niger*

b) Penecillium

c) Rizopus

d) Gloeosporium

78. Harvesting of tomatoes are done on the basis of

a) Color

b) Sugar and acid ratio

c) Oil

d) a and b

79. Post-harvest losses are

a) Physiological

b) Quantitative

c) Qualitative

d) All of the above

80. Post-harvest losses is due to the

a) Mechanical damage

b) Physiological deterioration

c) Disease, insect and pest

d) a, b and c

81. Stages of post-harvest losses are

a) Harvesting

b) Transportation/storage

c) Packaging

d) All of the above

82. True teas comes from which portion of plant

a) Stem

b) Tuber

c) Leaves

d) Beans

83. Which chemical is used to minimize browning in drying of fruits

a) Carbon dioxide

b) Chlorophyll

c) Sulphur dioxide

d) Benzene

84. Best time for harvesting of fruits in a day

a) Morning

b) Early morning

c) Evening

d) After noon

85. Which one is harvested by hand

a) Apple

b) Tomato

c) Citrus

d) All of the above

86. The main type of mechanical damage is

a) Compression/rubbing
b) Cuts
c) Impacts
d) All of the above

87. Principle method to dehydrate coffee bean extract is

a) Spray drying
b) Drum drying
c) Tunnel drying
d) All of the above

88. Ripening of fruits requires

a) CO_2
b) Hormone
c) Enzymes
d) oxygen

89. Major objective of vegetablc blanching is

a) Fixation of color
b) Inactivation of bacteria
c) Inactivation of enzymes
d) Removal of tissue gas

90. Causative spoilage organisms of dried fruits and vegetables are

a) Bacteria
b) Mold
c) Yeast
d) All of the above

91. Bacterial growth is check when water activity is less than the

a) 0.6
b) 0.7
c) 0.8
d) 0.9

92. During mango pickle preparation fermentation takes place

a) Alcoholic
b) Lactic acid
c) Acetic acid
d) Butyric acid

93. Potassium sorbate in fruits and vegetables preservation is most effective against

a) Bacteria
b) Yeasts
c) Molds
d) b and c

94. Organic acid as a preservative effective against

a) Lipolysis
b) Rancidity
c) Autolysis
d) Putrefaction

95. Storage life of fruits and vegetables is extended by keeping them in an atmosphere

a) High in CO_2 and low in O_2
b) High in O_2 and low in CO_2
c) At low temperature
d) At high relative humidity

96. Fruit flavor is a complex of

a) Taste and color
b) Taste and appearance
c) Taste and aroma
d) All of the above

97. Plant hormone that plays a key role in the ripening and senescence of fruits and vegetables is

a) Auxins
b) Ethylene
c) Papain
d) None of the above

98. Natural dormancy period in

a) Root crops
b) Tuber crops
c) Bulb crops
d) All of the above

99. Canning of vegetables is commonly done in

a) Plastic
b) Aluminum
c) Glass
d) Tin plate

100. The most suitable method for blanching of vegetables is

a) Steam
b) Hot water
c) Hot air
d) Microwave

101. Temperature, concentration of CO_2 and oxygen are the main environmental factors which influence the

a) Rate of respiration
b) Yield
c) Rate of growth of fruits & vegetables
d) All of the above

102. Fruit which is injured by exposure to temperatures less than 11 ^{0}C is

a) Banana
b) Apple
c) Pear
d) Orange

103. The main environmental factor to determine the rate of respiration is

a) O_2
b) CO_2
c) Relative humidity
d) Temperature

104. Which toxin produced during aerobic respiration of plant tissues that kills the cells if not removed are

a) Phytate
b) Formaldehyde
c) Ethyl alcohol and acetaldehyde
d) None of the above

105. Bitter pit in apple due to deficiency of

a) Nitrogen
b) Potassium
c) Calcium
d) Iron

106. Soft rot in citrus fruit is due to

a) Aspergilous
b) Mucor
c) Rhizopus
d) Pencillium

107. Citrus fruit harvesting is based on

a) Oil percentage
b) Color
c) Iodine
d) Sugar and acid ratio

108. Relative humidity for the storage of fruits is

a) Less than 70 per cent
b) 70-80 per cent
c) 85-95 per cent
d) 90-100 per cent

109. In CA storage oxygen concentration is reduced from 21 per cent to

a) 3 per cent
b) 7 per cent
c) 10 per cent
d) 15 per cent

110. Eutectic temperature is the temperature at which product

a) Above to freeze
b) About to freeze
c) No more freezable water is left
d) None of the above

111. Sparkling water consists of carbonated distilled water and salts in which CO_2 gas is added at the rate of

a) 0.1-0.2 volume
b) 0.2-0.3 volume
c) 0.7-0.8 volume
d) 4-5 volume

112. Ion exchange resins are not preferred for water purification because

a) Inefficiency
b) High cost
c) Less sensitivity
d) All of the above

113. Packaging materials for cut flowers is

a) Wooden boxes
b) Cardboard boxes
c) Plastic boxes
d) Caretes

114. Ethylene emission during transportation of cut flowers cause a disorder which called as

a) Bent neck
b) Bud opening
c) Sleepiness
d) Calyx splitting

115. First commodity for which grading and marketing rules were framed is

a) Mango
b) Tomato
c) Onion
d) Grape

116. Precursor of ethylene is

a) ABA
b) IAA
c) Tryptophane
d) Methionine

117. Cauliflowers curd can be stored for a month at

a) 15 °C with 60-65 per cent RH
b) 0 °C with 85-90 per cent RH
c) 15 °C with 60-80 per cent RH
d) 30 °C with 80 per cent RH

118. Elephant Foot Yam is the rich source of Vitamin

a) A and B
b) B and C
c) C and D
d) Only B

119. Tomato fruits are harvested for canning is at

a) Mature green stage
b) Immature green stage
c) Red ripe stage
d) Half-ripe/pink stage

120. Sprouting inhibitor for onion is

a) GA
b) Ethylene
c) Maleic Hydrazide (MH)
d) All of the above

121. Melons are picked for distant market at the stage of

a) Full-slip stage
b) Half-slip stage
c) Mature green stage
d) None of the above

122. For distant marketing tomato are harvested at

a) Mature green stage
b) Immature green stage
c) Red ripe stage
d) Half-ripe/pink stage

123. For low sugar content potatoes are stored at

a) 5 °C
b) 10 °C
c) 15 °C
d) 20 °C

124. For long term storage potato are stored at

a) 0-5 °C
b) 5-10 °C
c) 10-15 °C
d) 15-20 °C

125. For long-term storage of cucumber temperature are maintained at

a) 10 °C
b) 20 °C
c) 30 °C
d) None of the above

126. The limiting amino acid in green vegetables is

a) Lysine
b) Arginine
c) Methionine
d) Tryptophan

127. Staple vegetable is

a) Potato
b) Cauliflower
c) Chilli
d) Okra

128. Common refrigerant used in cold storage in our country

a) Ethylene
b) Carbide
c) Sodium benzoate
d) Ammonia

129. Mango variety with strong flavour is

a) Sindhu
b) Fazli
c) Langra
d) Dashahari

130. Canning variety of pineapple is

a) Mauritious
b) Queen
c) Kew
d) Cyenne

131. Richest source of riboflavin is

a) Papaya
b) Mango
c) Karonda
d) Bael

132. Richest source of iron is

a) Mango
b) Pomegranate
c) Bael
d) Dry karonda

133. Example of climacteric fruit is

a) Litchi
b) Citrus
c) Grape
d) Banana

134. Which of the following methods are generally not used in PHT

a) Filtering
b) Chilling
c) Fermenting
d) Boiling and drying

135. Cashew nut are harvested in the month of

a) February- May
b) May-June
c) August-September
d) July-August

136. Pernicious anemia occurs due to deficiency of

a) Vitamin B_1
b) Vitamin B_2
c) Vitamin B_5
d) Vitamin B_{12}

137. Drained weight of canned whole tomato is

a) 40 per cent
b) 50 per cent
c) 60 per cent
d) 70 per cent

138. Which of the following produce is being preserved by modified atmosphere packaging?

a) Flavored milk
b) Canned tomatoes
c) Prepackaged salad
d) Probiotic milk

139. Glycemic index refers to

a) Measure of protein levels in blood
b) Measure of glucose levels in blood
c) Measure of glucose levels in urine
d) b and c

140. The density that distinguishes HDPE from LDPE is

a) 0.58
b) 0.67
c) 0.87
d) 0.94

141. A promising nitrogen containing edible film is

a) Chitosan
b) Bees wax
c) Glucono lactone
d) Carnauba wax

142. Harvest Watch Technology is based on the principle of

a) Respiratory quotient
b) Ethylene flux
c) Chlorophyll fluorescence
d) Chemical luminescence

143. The major pigment of red capsicum is

a) B-carotene
b) Lycopene
c) Lutein
d) Capsanthin

144. T.S.S. may be define as

a) Toxic Shock Syndrome
b) Total System Service
c) Total Soluble Solids
d) Total Soluble Solid

145. Father of canning Industry is

a) Peter Decresenji
b) Johansberg
c) Burns and Prayag
d) M. Nicholus Appert

146. Fruits and vegetable processing was first started in the year

a) 1927 **b) 1857**

c) 1937 d) 1957

147. The book in titled "The art of preserving animal and vegetable substances for many years" was written by

a) Augustus de Heine **b) M. Nicholus Appert**

c) Papin d) Peter Durand

148. Energy required for heavy worker is

a) 80-180 kcal/hr b) 170-240 kcal/hr

c) 250-350 kcal/hr d) 350-500 kcal/hr

149. Coefficient of digestibility for fat is

a) 0.98 **b) 0.95**

c) 0.92 d) 0.85

150. Beer yeast is

a) *Saccharomyces cerevisiae* b) *Saccharomyces ellipsoideus*

c) *Saccharomyces carlsbergensis* d) *Schizosaccharomyces pombe*

151. Red colour of the tomato is due to

a) Anthocyanin b) Xanthophyll

c) Lycopene d) Carotene

152. In onion pink colour is due to

a) Anthocyanin b) Carotene

c) Xanthophyll **d) Quercitin**

153. Zero energy cool chamber is developed by

a) M K Rai and RN Singh **b) SK Roy and DS Khurdiya**

c) RP Roy and DK Khurana d) None of these

154. Hen and chiken disorder is associated with

a) Mango b) Tomato

c) Grapes d) Guava

155. The membrane lipid hypothesis is given by

a) Kidd and West b) James Harrison

c) Raison and Lyons d) Graham and Patterson

156. The point at which the dried products just become lumpy is known as

a) Danger Point
b) Saturated Point
c) Critical Point
d) Safety Point

157. A Point which have 5 per cent Low RH than the Critical Point is

a) Critical Point
b) Danger Point
c) Saturated Point
d) Safety Point

158. Fungus which mostly grown on grapes

a) Geotrichum
b) Penicillium
c) Botrytis
d) Colletotrichum

159. Vitamin D is chemically known as

a) Retinol
b) Cabalanin
c) Calciferol
d) Tocopherol

160. Lye peeling is done at temperature of

a) 75°C
b) 84°C
c) 93°C
d) 105°C

161. Which of the following is associated with 'browning' disorder

a) Apple
b) Cabbage
c) Cauliflower
d) Citrus

162. The membrane lipid hypothesis is given by

a) Kidd and West
b) Graham and Patterson
c) Raison and Lyons
d) James Harrison

163. What is the threshold level of ethylene in fruit and vegetable

a) 0.01 µL/L
b) 0.02 µL/L
c) 0.03 µL/L
d) 0.04 µL/L

164. Which of the following is a rapid precooling method

a) Forced air Cooling
b) Hydro Cooling
c) Vacuum Cooling
d) Evaporative Cooling

165. In cucumber, chilling- injury symptoms are occurred at

a) <7°C
b) 7°C
c) 10°C
d) >10°C

166. Degreening is not applicable in

a) Banana
b) Guava
c) Mango
d) Citrus

167. Which of the following is associated with 'browning' disorder

a) Apple **b) Cauliflower**

c) Cabbage d) Knol khol

168. Which vegetable is leading in processing

a) Potato **b) Tomato**

c) Watermelon d) Pumpkin

169. Breaker stage in tomato contain lycopene is

a) 10 per cent b) 15 per cent

c) 20 per cent d) 25 per cent

170. Which var. of tomato is suitable for processing?

a) Naveen b) Minakshi

c) Avinash **d) Pusa Uphar**

171. Chilli var. ______ suitable for colour extraction

a) Arka Abir b) Bullnose

c) Sweet Banana d) Indira

172. Export standard of okra

a) 6-8 cm b) 8-10 cm

c) 4-6 cm d) 10-12 cm

173. Dehydration ratio in onion is

a) 10: 1 b) 15: 1

c) 100: 1 d) 150: 1

174. Diameter of exportable garlic which contain 10-15 cloves

a) 30-40 cm b) 40-60 cm

c) 60-80 cm d) 80-100 cm

175. Maturity of pea is measured by

a) Lysometer b) Speedometer

c) Tendrometer d) Brix

176. Fuyu a product prepared from

a) Elephant foot yam b) Cassava

c) Yam d) Colocasia

177. which vegetables used as blood purifier

a) Elephant foot yam b) Cassava

c) Tomato d) Chilli

178. Sauerkraut are prepared from which type of cabbage

a) Red Cabbage
b) White Cabbage
c) Okra
d) None of these

179. ICMR recommends use of how much vegetable per day per capita

a) 300 g
b) 180 g
c) 250 g
d) 200 g

180. Sago is the important food product, which is derived from

a) Sweet Potato
b) Yam
c) Cassava
d) All of the above

181. TSS of ripe muskmelon lies between

a) 11-17 per cent
b) 17-20 per cent
c) 10-12 per cent
d) 20-25 per cent

182. Yerusseri product is prepared from which vegetables

a) Cucumber
b) Pumpkin
c) Radish
d) Carrot

183. Which refrigerant is commonly used is used in cold storage

a) Ethylene
b) Ammonia
c) Carbide
d) Sodium Bcnzoate

184. "Most suitable packaging material" for cut flowers is

a) Wooden boxes
b) Plastic boxes
c) Cardboard boxes
d) Carats

185. Emission of Ethylene during transportation of cut flowers cause a disorder which called as:

a) Bud opening
b) Sleepiness
c) Bent neck
d) Calyx splitting

186. Which is the precursor of Ethylene?

a) Tryptophane
b) Methionine
c) ABA
d) IAA

187. Cauliflower curds can be stored for a month at

a) 0°C with 85-90 per cent RH
b) 15°C with 60-80 per cent RH
c) 15°C with 60-65 per cent RH
d) 20 °C with 50-70 per cent RH

188. "Elephant's Foot Yam" is rich source of Vitamin

a) A and B
b) B and C
c) C and D
d) Only B

189. Tomato fruits for canning are harvested at

a) Mature green stage
b) Immature green stage
c) Red ripe stage
d) Half-ripe/pink stage

190. Which chemical is used for controlling sprouting of onions in storage?

a) Maleic Hydrazide (MH)
b) Ethylene (C_2H_4)
c) GA3
d) All of these

191. Melons for distant marketing arc picked at

a) Half-slip stage
b) Full-slip stage
c) Green mature stage
d) None of the above

192. For Low Sugar content, potato tubers are stored at

a) 5 °C
b) 10 °C
c) 15 °C
d) 20 °C

193. Which bean is used for extraction of gum?

a) Broad bean
b) Cluster bean
c) French bean
d) Hyacinth bean

194. Yellow-coloured vegetables are rich source

a) Vitamin A
b) Vitamin B
c) Vitamin C
d) Vitamin D

195. Enzyme responsible for converting proto-pectin into pectin is

a) PME
b) Proto-pectinase
c) Poly Galucturonase
d) Pecfmase

196. Storage temperature for Asparagus is

a) 0 – 5 °C
b) 5 – 7 °C
c) 7-11 °C
d) 10-15 °C

197. Moisture content in dried vegetable is

a) 2 per cent
b) 3 per cent
c) 5 per cent
d) 6 per cent

198. Bacteria which is used to absorb ethylene from the storage chamber is

a) Agrobacterium
b) Mycobacterium
c) Bacillus
d) Azotobacter

199. Toughening effect on canned bean is due to

a) K

b) Ca

c) S

d) None of the above

200. Leaf change is an important maturity index for

a) Bulbous vegetables

b) Seed vegetables

c) Cucurbits

d) Leafy vegetables

One Liner in PHT

- ☆ Nicolas Apart is known as the father of food preservation.
- ☆ A heat treatment food material at 72 ^{0}C for 15 seconds, or 63 ^{0}C for 30 minutes: pasteurization
- ☆ Juices are mostly preserved by freezing
- ☆ Original colour of beverages a for longer period are retained by Benzoic acid
- ☆ Removal of moisture from food materials for preservation is known as dehydration
- ☆ Killing of microorganisms in food by heat processing
- ☆ The concentration of sugar required for the preservation of fruit jam is 66 per cent
- ☆ The concentration of salt sufficient to preserve most of the food products is 15-25 per cent
- ☆ A thermal process mostly used vegetables able prior to freezing, drying or canning in order to soften the texture is blanching
- ☆ The solution made by dissolving salt (NaOH) in water is called as brine solution
- ☆ 10 per cent fruit juice and 10 per cent soluble solids contains in ready-to-serve
- ☆ Nectar is a type of fruit beverage prepared from Aonla, Bael, Mango, Papaya
- ☆ Squash contains 25 per cent juice and 40 per cent TSS
- ☆ Mango pulp is preserved by sugar
- ☆ Squash and syrup are diluted before serving
- ☆ TSS of jam should be 68 per cent
- ☆ TSS content in fruit juice is measured by hand or digital refractometer
- ☆ A semi-solid transparent product prepared from pectin containing fruit is jelly
- ☆ For jelly making fruit should be harvested at a firm ripe stage
- ☆ pH of the final jelly should be 3.2
- ☆ The instrument used to check the pectin content is jellimeter
- ☆ A fruit vegetable impregnated with cane sugar or glucose syrup is known as candied

- ☆ A product prepared from the strained pulp of fully ripe tomato fruits after cooking is known as tomato sauce or ketchup
- ☆ Salt concentration in the pickle is maintained at 8-10 per cent
- ☆ CFTRI denotes Central Food Technological Research Institute, Mysore, Karnataka
- ☆ Potassium permanganate is used in sealed packets to reduce the level of ethylene
- ☆ Freezing preservation involves the principle of removal of heat
- ☆ Single-cell protein produced by organism *E. coli*
- ☆ The ideal temperature for ripening in mango is 20-25 ^{0}C and 80-85 per cent RH
- ☆ The most accepted pathway of ethylene biosynthesis is the methionine pathway
- ☆ Conversion of glucose to pyruvate by EMP pathway takes place in the nucleolus
- ☆ The important quality parameter for export of dehydrated onion is white colour with low pungency
- ☆ Waxing in fruits and vegetables is mainly done to reduce the transpiration
- ☆ A visible shrinkage of freshly harvested fruits and vegetables appears when the moisture loss is around 10-15 per cent
- ☆ Lye peeling is a technique for the peeling of fruits and vegetables using 1-2 per cent NaOH (caustic Soda) solution
- ☆ Exhausting term in canning is associated with the removal of air
- ☆ A dividing line of pH between acid and non-acid food is 4.6
- ☆ In food preservation the most commonly used irradiation is gamma rays
- ☆ The natural red colour pigment which changes its colour with pH is anthocyanin
- ☆ Hedonic scale is used to determine the sensory quality
- ☆ Sauerkraut means acid cabbage
- ☆ For preparing fruit jelly, the quantity of sugar to be added depends upon the pectin content
- ☆ Oxidation of fats in potato chips is prevented by N_2 gas packaging
- ☆ Spongy tissues in mango is due to the physiological disorder
- ☆ The browning of fresh potato cut surface is due to enzymatic browning
- ☆ Vacuum cooling is ideally suitable for leafy vegetables
- ☆ M. Nicholas Appert is known as the father of the canning industry

- ☆ The flat sour spoilage-causing bacteria in canned foods are thermophilic
- ☆ Formation of brown mass is due to the burning of dry sugar is called as browning
- ☆ The food safety act was issued by MoFPI in the year 2006
- ☆ Covering candied fruits and vegetables with a thin transparent coating of sugar which imparts them glossy appearance is known as glazing
- ☆ Green portion or greening in potato is undesirable as it may contain harmful substance like solanin
- ☆ Pro-Vitamin A activity found in βcarotene
- ☆ Name of the lowest water activity value for permitting the growth of osmophilic yeasts is 0.61
- ☆ The first example of the use of bio-agent for controlling post-harvest diseases is the use of *Trichoderma harzianum*
- ☆ Symptoms associated with chilling injury are freeze burn and discolouration
- ☆ Laminated film with the lowest water vapour and oxygen transmission rate is PE
- ☆ Taste buds for bitterness are located at basel portion or back of the tongue
- ☆ Preservative is mainly used in baked products is sorbic acid
- ☆ Which chilli has the maximum SHU units in the world Bhut Jaulokia

Match the Following

Q1.			
I.	Dehydration	**a)**	Pectin
II.	Climacteric fruit	**b)**	Storage structure
III.	Jelly	**c)**	Pigment
IV.	Cool chamber	**d)**	Drying under controlled condition
V.	Chlorophyll	**e)**	Mango

Ans.				
I.	**II.**	**III.**	**IV.**	**V.**
d)	**e)**	**a)**	**b)**	**c)**

Q2.			
I.	Invertase	**a)**	Pineapple
II.	Papain	**b)**	Starch
III.	Lipase	**c)**	Sugar
IV.	Diastase	**d)**	Fat
V.	Bromolin	**e)**	Papaya

Ans.				
I.	**II.**	**III.**	**IV.**	**V.**
c)	**e)**	**d)**	**b)**	**a)**

Q3.			
I.	Tomato ketchup	**a)**	Ascorbic acid
II.	Anti-browning agent	**b)**	Osmosis
III.	Cell shrinkage	**c)**	Sodium benzoate
IV.	Heat unit	**d)**	Citrus fruit
V.	Dieback	**e)**	Peas

Ans.				
I.	**II.**	**III.**	**IV.**	**V.**
c)	**a)**	**b)**	**e)**	**d)**

Q4.			
I.	Evaporative cooling	**a)**	Cold sterilization
II.	Elevated CO_2	**b)**	Food safety
III.	Irradiation	**c)**	Controlled atmosphere
IV.	VHT	**d)**	Quarantine
V.	HACCP	**e)**	Pusa zero-energy cool chamber

Ans.				
I.	**II.**	**III.**	**IV.**	**V.**
e)	**c)**	**a)**	**d)**	**b)**

Q5.			
I.	Hurdle concept	**a)**	Dates
II.	Aseptic processing	**b)**	Prepacked salads
III.	Vacuum packaging	**c)**	Potato chips
IV.	Modified atmosphere packaging	**d)**	Tetra Pak
V.	Extrusion processing	**e)**	Corn puffs

Ans.				
I.	**II.**	**III.**	**IV.**	**V.**
b)	**d)**	**c)**	**e)**	**a)**

Q6.			
I.	CIPC	**a)**	Sprout inhibitor
II.	Semper fresh	**b)**	Fungicide
III.	Bavistin	**c)**	Preservative
IV.	Potassium sorbate	**d)**	Surface coating
V.	1-MCP	**e)**	Ethylene inhibitor

Ans.				
I.	**II.**	**III.**	**IV.**	**V.**
a)	**d)**	**b)**	**c)**	**e)**

Q7.			
I.	Apiin	**a)**	Yams
II.	Calcium oxalate	**b)**	Cassava
III.	CN glycosides	**c)**	Cucurbits
IV.	Cucurbitacins	**d)**	Celery
V.	Dioscorine	**e)**	Colocasia

Ans.				
I.	II.	III.	IV.	V.
d)	e)	d)	c)	a)

Q8.			
I.	Haemaglutine	a)	Amaranthus
II.	Oxalic acid	b)	French bean
III.	Saponine	c)	Brinjal
IV.	Sinigrin	d)	Tomato, Asparagus
V.	Solasodine	e)	Cole crops

Ans.				
I.	II.	III.	IV.	V.
b)	a)	d)	e)	c)

Q9.			
I.	Trypsin inhibitor	a)	CIPC
II.	Sprout inhibitor	b)	Brinjal
III.	Serotonin	c)	Cole crops
IV.	Solasodine	d)	Peas
V.	Sinigrin	e)	Water melon

Ans.				
I.	II.	III.	IV.	V.
d)	a)	e)	b)	c)

Q10.			
I.	Tomato	a)	Tapioca
II.	Brinjal	b)	Wholesome food
III.	Pumpkin	c)	Vilayati baigan
IV.	Cassava	d)	Aubergine
V.	Musk melon	e)	Vegetables of immense value

Ans.				
I.	II.	III.	IV.	V.
c)	d)	e)	a)	b)

Q11.			
I.	Lima bean	**a)**	Butter bean
II.	French bean	**b)**	Faba bean
III.	Dolichus bean	**c)**	Goa bean, four-angled bean
IV.	Broad bean	**d)**	Kidney bean
V.	Winged bean	**e)**	Indian bean

Ans.				
I.	**II.**	**III.**	**IV.**	**V.**
a)	**d)**	**e)**	**b)**	**c)**

Q12.			
I.	Capsicum	**a)**	White gourd, wax gourd
II.	Drumstick	**b)**	Southern pea
III.	Cowpea	**c)**	Chappan kaddu
IV.	Ash gourd	**d)**	Bell pepper, sweet pepper
V.	Summer squash	**e)**	Horse radish tree, Sahijan

Ans.				
I.	**II.**	**III.**	**IV.**	**V.**
d)	**e)**	**b)**	**a)**	**c)**

Q13.			
I.	Potato	**a)**	0-1.7 ^{0}C
II.	Brinjal	**b)**	3-4 ^{0}C
III.	Garlic	**c)**	4 ^{0}C
IV.	Cole crops	**d)**	0 ^{0}C
V.	Sprouting broccoli	**e)**	10-11 ^{0}C

Ans.				
I.	**II.**	**III.**	**IV.**	**V.**
b)	**e)**	**d)**	**a)**	**c)**

Q14.			
I.	Onion, garlic	**a)**	Dithiolthiones
II.	Cole crops	**b)**	Diosgenin
III.	Bitter gourd	**c)**	3-n-butyl pthalide
IV.	Celery	**d)**	Sulphur compounds
V.	Yams	**e)**	Cheratin

Ans.				
I.	**II.**	**III.**	**IV.**	**V.**
d)	**a)**	**e)**	**c)**	**b)**

Q15.			
I.	Onion, garlic	**a)**	Indole-3-carbinol
II.	Bitter gourd	**b)**	Glucoraphinanin
III.	Cabbage	**c)**	Lutein
IV.	Broccoli	**d)**	Quercetin
V.	Lettuce	**e)**	Cheratin

Ans.				
I.	**II.**	**III.**	**IV.**	**V.**
d)	**e)**	**a)**	**b)**	**c)**

Q16.			
I.	Pusa Rudhira	**a)**	Anthocyanin
II.	Pusa Rohini	**b)**	Lycopene
III.	Palam Hriday	**c)**	β-carotene
IV.	Arka Chandan	**d)**	Capsanthin
V.	KTPL-9	**e)**	Lycopene

Ans.				
I.	**II.**	**III.**	**IV.**	**V.**
b)	**e)**	**a)**	**c)**	**d)**

Q17.			
I.	Serotonin	**a)**	Potato
II.	Solasodine	**b)**	Yam
III.	Salanin	**c)**	Tomato
IV.	Saponine	**d)**	Water melon
V.	Dioscorine	**e)**	Brinjal

Ans.				
I.	**II.**	**III.**	**IV.**	**V.**
d)	**e)**	**a)**	**c)**	**b)**

Q18.			
I.	Cauliflower	**a)**	Corm
II.	Cabbage	**b)**	Fruits
III.	Elephant Foot Yam	**c)**	Prefloral apical meristem
IV.	Tomato, Brinjal	**d)**	Potato
V.	Modified stem	**e)**	Head

Ans.				
I.	**II.**	**III.**	**IV.**	**V.**
c)	**e)**	**a)**	**b)**	**d)**

Q19. Effect of ethylene			
I.	Hard core	**a)**	Lettuce
II.	Fibrousness	**b)**	Cucumber
III.	Russet sprouting, discoloration	**c)**	Asparagus
IV.	Degreening	**d)**	Sweet potato
V.	Celery	**e)**	Blanching

Ans.				
I.	**II.**	**III.**	**IV.**	**V.**
d)	**c)**	**a)**	**b)**	**e)**

Q20.			
I.	Flower bud	**a)**	Onion
II.	Swollen inflorescence	**b)**	Artichoke
III.	Leaf blade	**c)**	Potato
IV.	Bulb	**d)**	Broccoli
V.	Stem tuber	**e)**	Spinach

Ans.				
I.	**II.**	**III.**	**IV.**	**V.**
b)	**d)**	**e)**	**a)**	**c)**

Q21. Aroma compounds			
I.	Cucumber-raw	**a)**	2-Mthoxy-3-ethyl pyrazine, 2,5-Dimethyl pyrazine
II.	Cucumber-cooked	**b)**	4-Methylthio-*trans*-3-butenyl isothiocynate

III.	Mushroom	**c)**	2, 6-Nonadienal
IV.	Potato	**d)**	1-Octen-3-01, lenthionine
V.	Radish	**e)**	Dimethyl disulphide

Ans.				
I.	**II.**	**III.**	**IV.**	**V.**
c)	**e)**	**d)**	**a)**	**b)**

Q22. Highest content			
I.	Bathua leaves	**a)**	Protein
II.	Drumstick leaves	**b)**	Fenugreek
III.	Lima bean	**c)**	Carbohydrates
IV.	Tapioca	**d)**	Vitamin A
V.	Drumstick leaves	**e)**	Vitamin C

Ans.				
I.	**II.**	**III.**	**IV.**	**V.**
d)	**b)**	**a)**	**c)**	**e)**

Q23. Highest content			
I.	Amaranthus	**a)**	Calcium
II.	Spinach	**b)**	Potassium
III.	Agathi	**c)**	Calorific value
IV.	Potato	**d)**	Fibre
V.	Tapioca	**e)**	Iron

Ans.				
I.	**II.**	**III.**	**IV.**	**V.**
e)	**b)**	**a)**	**d)**	**c)**

Q24.			
I.	TPS	**a)**	C. M. Rick
II.	SPT	**b)**	Dr. Harbhajan Singh
III.	Tomato	**c)**	Dr. S. Ramanujam
IV.	Okra	**d)**	Dr. S. P. Singh
V.	Bottle gourd	**e)**	Dr. Pushkarnath

Ans.				
I.	**II.**	**III.**	**IV.**	**V.**
c)	**e)**	**a)**	**b)**	**d)**

Q25. Important chemical content			
I.	Linalool	**a)**	Fenugreek
II.	Cuminol	**b)**	Coriander
III.	Diosgenin	**c)**	Clove
IV.	Terbein	**d)**	Nutmeg
V.	Eugenol	**e)**	Cumin

Ans.				
I.	**II.**	**III.**	**IV.**	**V.**
b)	**e)**	**a)**	**d)**	**c)**